THE HUMAN WORTH
OF RIGOROUS THINKING

THE HUMAN WORTH
OF RIGOROUS THINKING

ESSAYS AND ADDRESSES

BY

CASSIUS J. KEYSER, *1862 - 1947*

SECOND EDITION, ENLARGED

Essay Index Reprint Series

 BOOKS FOR LIBRARIES PRESS
FREEPORT, NEW YORK

INTERNATIONAL STANDARD BOOK NUMBER:
0-8369-2169-0

LIBRARY OF CONGRESS CATALOG CARD NUMBER:
71-142651

PRINTED IN THE UNITED STATES OF AMERICA

PREFACE

THE following fifteen essays and addresses have appeared, in the course of the last fifteen years, as articles in various scientific, literary, and philosophical journals. For permission to reprint I have to thank the editors and managers of *The Columbia University Quarterly*, *The Columbia University Press*, *Science*, *The Educational Review*, *The Bookman*, *The Monist*, *The Hibbert Journal*, and *The Journal of Philosophy, Psychology and Scientific Methods*.

The title of the volume indicates its subject. The fact that one of the essays, the initial one, bears the same title is hardly more than a mere coincidence, for all of the discussions deal with the subject in question and nearly all of them deal with it directly, consciously, and in terms.

In passing from essay to essay the attentive reader will notice a few repetitions of thought and possibly a few in forms of expression. Such reiterations, which owe their presence to the occasional character of the essays and to the aims and circumstances that originally controlled their composition, may, it is hoped, be regarded by the charitable reader less as blemishes than as means of emphasizing important considerations.

CASSIUS J. KEYSER.

April 14, 1916.

PREFACE TO THE SECOND EDITION

NATURALLY I am gratified to know that the demand for *The Human Worth of Rigorous Thinking* (1916) has recently increased rapidly. As the demand now exceeds the supply it seems desirable to reprint the volume or else to publish a new edition of it. The latter alternative has seemed best.

The present (second) edition differs from the first one in the following respects: some errors of type have been corrected and a few statements have been rephrased in the interest of clarity; various notes have been added, most of them referring to the most recent literature dealing with cardinal matters presented in the text; to the fifteen essays and addresses composing the original volume have been added two new addresses, one of them entitled *Man and Men* and the other one *Educational Ideals That are Most Worthy of Loyalty;* finally, the volume includes an index of all the authors mentioned in course of the discussions.

It occasionally happens that a reader writes to an author asking for further light or offering criticism and sometimes a word of praise or blame. Experience has taught me to value such communications as sometimes leading to helpful correspondence and not infrequently to the discovery of spiritual comradeship.

CASSIUS J. KEYSER

MAY 22, 1925

CONTENTS

THE HUMAN WORTH OF RIGOROUS THINKING [1]

But in the strong recess of Harmony
Established firm abides the rounded Sphere.
— EMPEDOCLES

NEXT to the peaceful pleasure of meeting genuine curiosity, half-way, upon its own ground, comes the joy of combat when an attack upon some valued right or precious interest of the human spirit requires to be repelled. Indeed, given a competent jury, hardly any other undertaking could be more stimulating than to defend mathematics from a charge of being unworthy to occupy, in the hierarchy of arts and sciences, the high place to which, from the earliest times, the judgment of mankind has assigned it.[2] But, unfortunately, no such accusation has been brought, brought, that is, by persons of such scientific qualifications as to give their opinion in the premises weight enough to call for serious consideration. Mathematics has been often praised by the scientifically incompetent; it has not, so far as I am aware, been dispraised, or its worth challenged or denied, by the scientifically competent. The age-long immunity of mathematics from authoritative arraignment, and the high esti-

¹ An address delivered before the Mathematical Colloquium of Columbia University, October 13, 1913. Printed, with slight change, in *Science,* December 5, 1913; also, with other slight changes, printed in *The Columbia University Quarterly,* June, 1914, under the title "The Study of Mathematics." The substance of the address was delivered before the mathematics section of the California High School Teachers Association, August, 1915, at Berkeley, California.

² See Moritz's magnificent *Memorabilia Mathematica,* The Macmillan Company.

mation in which the science has been almost universally held in enlightened times and places, unite to give it a position nearly, if not quite, unique in the history of criticism. Perhaps it were better not so. Mathematicians have a sense of security to which, it may be, they are not entitled in a critical age and a reeling world. Conceivably it might have been to the advantage of mathematics and not only of mathematics but of science in general, of philosophy, too, and the general enlightenment, if in course of the centuries mathematicians had been now and then really compelled by adverse criticism of their science to discover and to present not only to themselves but acceptably to their fellow-men the deeper justifications, if such there be, of the world's approval and applause of their work. However that may be, no one is likely to dissent from the opinion of Mr. Bertrand Russell that " in regard to every form of activity it is necessary that the question should be asked from time to time: what is its purpose and ideal?   In what way does it contribute to the beauty of human existence? "   An inquiry that is thus necessary for the general welfare ought to be felt as a duty, unless, more fortunately, it be • felt as a pleasure.

Why study mathematics?   What are the rightful claims of the science to human regard?   What are the grounds upon which a university may justify the annual expenditure of thirty to fifty thousands of dollars to provide for mathematical instruction and mathematical research?

A slight transformation of the questions will help to disclose their significance and may give a quicker sense of their poignancy and edge.   What is mathematics?   I hasten to say that I do not intend to detain the reader and thus perhaps to dampen his interest with a definition of mathematics, though it must be said that the discovery of what mathematics is, is doubtless one of the very great scientific achievements of the nineteenth century. The question asks, not for a definition of the science, but for a brief and helpful description of it — for an obvious mark or aspect of it that will enable us to know what it is that we are here writing or reading about. Well, mathematics may be viewed either as an enterprise or as a body of achievements. As an enterprise mathematics is characterized by its aim, and its aim is to think rigorously whatever is rigorously thinkable or whatever may become rigorously thinkable in course of the upward striving and refining evolution of ideas. As a body of achievements mathematics consists of all the results that have come, in the course of the centuries, from the prosecution of that enterprise: the truth discovered by it; the doctrines created by it; the influence of these, through their applications and their beauty, upon the advancement of civilization and the weal of man.

Our questions now stand: Why should a human being desire to share in that spiritual enterprise which has for its aim to think rigorously whatever is or may become rigorously thinkable and to " frame a world according to a rule of divine perfection " ? Why should men and women seek some knowledge of that variety of perfection with which men and women have enriched life and the world by rigorous thought? What are the just claims to human regard of perfect thought and the spirit of perfect thinking? Upon what grounds may a university justify the annual expenditure of thirty to fifty thousands of dollars to provide for the disciplining of men and women in the art of thinking rigorously and for the promotion of research in the realm of exact thought?

Such are the questions. They plainly sum themselves in one: among the human agencies that ameliorate life, what is the rôle of rigorous thinking? What is the rôle of the spirit that always aspires to the attainment of logical perfection?

Evidently that question is not one for adequate handling in a brief magazine article by an ordinary student of mathematics. Rather is it a subject for a long series of lectures by a learned professor of the history of civilization. Indeed so vast is the subject that even an ordinary student of mathematics can detect some of the more obvious tasks such a philosophic historian would have to perform and a few of the difficulties he would doubtless encounter. It may be worth while to mention some of them.

Certainly one of the tasks, and probably one of the difficulties also, would be that of securing an audience — an audience, I mean, capable of understanding the lectures, for is not a genuine auditor a listener who understands? To understand the lectures it would seem to be necessary to know what that is which the lectures are about — that is, it would be necessary to know what is meant by rigorous thinking. To know this, however, one must either have consciously done some rigorous thinking or else, at the very least, have examined some specimens of it pretty carefully, just as, in order to know¹ what good art is, it is, in general, essential either to have produced good art or to have attentively examined some specimens of it, or to have done both of these things. Here, then, at the outset our historian would meet a serious difficulty, unless his audience chanced to be one of mathematicians, which is (unfortunately) not likely, inasmuch as the great majority of mathematicians are so exclusively interested in mathematical study or teaching

or research as to be but little concerned with the philosophical question of the human worth of their science.
It is, therefore, easy to see how our lecturer would have
to begin.

Ladies and gentlemen, we have met, he would say, to
open a course of lectures dealing with the rôle of rigorous
thinking in the history of civilization. In order that the
course may be profitable to you, in order that it may be
a course in ideas and not merely or mainly a verbal
course, it is essential that you should know what rigorous
thinking is and what it is not. Even I, your speaker,
might reasonably be held to the obligation of knowng
that.

It is reasonable, ladies and gentlemen, to assume that in
the course of your education you neglected mathematics,
and it is therefore probable or indeed quite certain that,
notwithstanding your many accomplishments, you do not
quite know or rather, perhaps I should say, you are very
far from knowing what rigorous thinking is or what it is
not. Of course, it is, generally speaking, much easier to
tell what a thing is *not* than to tell what it is, and I might
proceed by way of a preliminary to indicate roughly what
rigorous thinking is not. Thus I might explain that
rigorous thinking, though much of it has been done in the
world and though it has produced a large literature, is
nevertheless a relatively rare phenomenon. I might point
out that a vast majority of mankind, a vast majority of
educated men and women, have not been disciplined to
think rigorously even those things that are most available
for such thinking. I might point out that, on the other
hand, most of the ideas with which men and women have
constantly to deal are as yet too nebulous and vague, too
little advanced in the course of their evolution, to be
available for concatenative thinking and rigorous dis-

course. I should have to say, he would add, that, on these accounts, most of the thinking done in the world in a given day, whether done by men in the street or by farmers or factory-hands or administrators or historians or physicians or lawyers or jurists or statesmen or philosophers or men of letters or students of natural science or even mathematicians (when not strictly employed in their own subject), comes far short of the demands and standards of rigorous thinking.

I might go on to caution you, our speaker would say, against the current fallacy — recently advanced by eloquent writers to the dignity of a philosophical tenet — of regarding what is called successful action as the touchstone of rigorous thinking. For you should know that much of what passes in the world for successful action proceeds from impulse or instinct and not from thinking of any kind; you should know that no action under the control of non-rigorous thinking can be strictly successful except by the favor of chance or through accidental compensation of errors; you should know that most of what passes for successful action — most of what the world applauds and even commemorates as successful action — so far from being really successful, varies from partial failure to failure that, if not total, would at all events be fatal in any universe that had the economic decency to forbid, under pain of death, the unlimited wasting of its resources. The dominant creature of such a universe would be, in fact, a superman. In our world the natural resources of life are superabundant, and man is poor in reason because he has been the prodigal son of a too opulent mother. But, ladies and gentlemen, you will know better what rigorous thinking is not when once you have learned what it is. This, however, cannot well be learned in a course of lectures in which that knowledge is presumed. I have, therefore, to adjourn this course until such time as you shall have gained that knowledge. It cannot be gained by reading about it or hearing about it. The easiest way — for most persons the only way — to gain it is to examine with exceeding patience and care some specimens, at least one specimen, of the literature in which rigorous thinking is embodied. Such a specimen is Dr. Thomas L. Heath's magnificent edition of Euclid, where an excellent translation of the Elements from the definitive text of Heiberg is set in the composite light of critical commentary from Aristotle down to the keenest logical microscopists and histologists of our own day. If you think Euclid too ancient or too stale even when seasoned with the wit of more than two thousand years of the acutest criticism, you may find a shorter and possibly a fresher way by examining minutely such a work as Veronese's *Grundzüge der Geometrie* or Hilbert's famous *Foundations of Geometry* or Peano's *Sui Numeri Irrazionali*. In works of this kind and not elsewhere you will find in its nakedness, purity, and spirit, what you have neglected and what you need. You will note that in the beginning of such a work there is found a system of assumptions or postulates, discovered the Lord only and a few men of genius know where or how, selected perhaps with reference to simplicity and clearness, certainly selected and tested with respect to their compatibility and independence, and, it may be, with respect also to categoricity. You will not fail to observe with the utmost minuteness, and from every possible angle, how it is that upon these postulates as a basis there is built up by a kind of divine masonry, little step by step, a stately structure of ideas, an imposing edifice of rigorous thought, a towering architecture of doctrine that is at once beautiful, austere, sublime, and eternal. Ladies and gentlemen, our speaker will say, to accomplish that examination will require twelve months of pretty assiduous application. The next lecture of this course will be given one year from date. On resuming the course what will our philosopher and historian proceed to say? He will begin to say what, if he says it concisely, will make up a very large volume. Room is lacking here, even if competence were not, for so much as an adequate outline of the matter. It is possible, however, to draw with confidence a few of the larger lines that such a sketch would have to contain.

What is it that our speaker will be obliged to deal with first? I do not mean obliged logically nor obliged by an orderly development of his subject. I mean obliged by the expectation of his hearers. Every one can answer that question. For presumably the audience represents the spirit of the times, and this age is, at least to a superficial observer, an age of engineering. Now, what is engineering? Well, the Charter of the Institution of Civil Engineers tells us that engineering is the " art of directing the great sources of power in Nature for the use and convenience of man." [3] By Nature here must be meant external or physical nature, for, if internal nature were also meant, *every* good form of activity would be a species of engineering, and maybe it is such, but that is a claim which even engineers would hardly make and poets would certainly deny. Use and convenience — these are the key-bearing words. It is perfectly evident that our lecturer will have to deal first of all with what the world would call the " utility " of rigorous thinking, that is to say, with the applications of mathematics and especially

---

[3] For a far nobler conception of engineering see Korzybski's *Manhood of Humanity* (Dutton & Co.). Also the chapter on Science and Engineering in Keyser's *Mathematical Philosophy* (Dutton).

with its applications to problems of engineering. If he really knows profoundly what mathematics is, he will not wish to begin with applications nor even to make applications a major theme of his discourse, but he must, and he will do so uncomplainingly as a concession to the external-mindedness of his time and his audience.

He will not only desire to show his audience applications of mathematics to engineering, but, being an historian of civilization, he will especially desire to show them the development of such applications from the earliest times, from the building of pyramids and the mensuration of land in ancient Egypt down to such splendid modern achievements as the designing and construction of an Eads Bridge, an ocean Imperator or a Panama Canal. The story will be long and difficult, but it will edify. The audience will be amazed at the truth if they understand. If they do not understand the truth fully, our speaker must at all events contrive that they shall see it in glimmers and gleams and, above all, that they shall acquire a feeling for it. They must be led to some acquaintance with the great engineering works of the world, past and present; they must be given an intelligible conception of the immeasurable contribution such works have made to the comfort, convenience, and power of man; and especially must they be convinced of the fact that, not only would the greatest of such achievements have been, except for mathematics, utterly impossible, but that such of the lesser ones as could have been wrought without mathematical help could not have been thus accomplished without wicked and pathetic waste both of material resources and of human toil. In respect to this latter point, the relation of mathematics to practical economy in large affairs, our speaker will no doubt invite his hearers to read and reflect upon the ancient work of Frontinus on the *Water Supply of the City of Rome* in order that thus they may gain a vivid idea of the fact that the most *practical* people of history, despising mathematics and the finer intellectualizations of the Greeks, were unable to accomplish their own great engineering feats except through appalling waste of materials and men. Our lecturer will not be content, however, with showing the service of mathematics in the prevention of waste; he will show that it is indispensable to the productivity and trade of the modern world. Before quitting this division of his subject he will have demonstrated that, if all the contributions which mathematics has made, and which nothing else could make, to navigation, to the building of railways, to the construction of ships, to the subjugation of wind and wave, electricity and heat, and many other forms and manifestations of energy, he will have demonstrated, I say, and the audience will finally understand, that, if all these contributions of mathematics were suddenly withdrawn, the life and body of industry and commerce would suddenly collapse as by a paralytic stroke, the now splendid outer tokens of material civilization would perish, and the face of our planet would quickly assume the aspect of a ruined and bankrupt world.

As our lecturer has been constrained by circumstances to back into his subject, as he has, that is, been compelled to treat first of the service that mathematics has rendered engineering, he will probably next speak of the applications of mathematics to the so-called natural sciences — the more properly called experimental sciences — of physics, chemistry, biology, economics, psychology, and the like. Here his task, if it is to be, as it ought to be, expository as well as narrative, will be exceedingly hard. For how can he weave into his narrative an intelligible exposition of Newton's *Principia,* Laplace's

Mécanique Céleste, Lagrange's *Mécanique Analytique,*
Gauss's *Theoria Motus Corporum Coelestium,* Fourier's
Théorie Analytique de la Chaleur, Maxwell's *Electricity
and Magnetism,* not to mention scores of other equally
difficult and hardly less important works of a mathemat-
ical-physical character? Even if our speaker knew it
all, which no man can, he could not tell it all intelligibly
to his hearers. These will have to be content with a
rather general and superficial view, with a somewhat
vague intuition of the truth, with fragmentary and ana-
logical insights gained through settings forth of great
things by small; and they will have to help themselves
and their speaker, too, by much pertinent reading. No
doubt the speaker will require his hearers, in order that
they may thus gain a tolerable perspective, to read well
not only the first two volumes of the magnificent work of
John Theodore Merz dealing with the *History of European Thought in the Nineteenth Century,* but also many
selected portions of the kindred literature there cited in
richest profusion. The work treats mainly of natural
science, but it deals with it philosophically, under the
larger aspect, that is, of science regarded as Thought.
By the help of such literature in the hands of his auditors,
our lecturer will be able to give them a pretty vivid sense
of the great and increasing rôle of mathematics in sug-
gesting, formulating, and solving problems in all branches
of natural science. Whether it be with " the astronomical view of nature " that he is dealing, or " the atomic
view " or " the mechanical view " or " the physical view "
or " the morphological view " or " the genetic view " or
" the vitalistic view " or " the psychophysical view "
or " the statistical view," in every case, in all these
great attempts of reason to create or to find a cosmos
amid the chaos of the external world, the presence of

mathematics and its manifold service, both as instrument and as norm, illustrate and confirm the Kantian and Riemannian conception of natural science as " the attempt to understand nature by means of exact concepts."

In connection with this division of his subject, our speaker will find it easy to enter more deeply into the spirit and marrow of it. He will be able to make it clear that there is a sense, a just and important sense, in which all thinkers and especially students of natural science, though their thinking is for the most part not rigorous, are yet themselves contributors to mathematics. I do not refer to the powerful stimulation of mathematics by natural science in furnishing it with many of its problems and in constantly seeking its aid. What I mean is that all thinkers and especially students of natural science are engaged, both consciously and unconsciously, both intentionally and unintentionally, in the mathematicization of concepts — that is to say, in so transforming and refining concepts as to fit them finally for the amenities of logic and the austerities of rigorous thinking. We are dealing here, our speaker will say, with a process transcending conscious design. We are dealing with a process deep in the nature and being of the psychic world. Like a child, an idea, once it is born, once it has come into the realm of spiritual light, possibly long before such birth, enters upon a career, a career, however, that, unlike the child's, seems to be immortal. In most cases and probably in all, an idea, on entering the world of consciousness, is vague, nebulous, formless, not at once betraying either what it is or what it is destined to become. Ideas, however, are under an impulse and law of amelioration. The path of their upward striving and evolution — often a long and winding way — leads towards precision and perfection of form. The goal is mathematics. Witness, for example,

the age-long travail and aspiration of the great concept
now known as mathematical *continuity* — a concept
whose inner structure is even now known and understood
only of mathematicians, though the ancient Greeks
helped in modeling its form and though it has long been,
if somewhat blindly, yet constantly employed in natural
science, as when a physicist, for example, or an astronomer uses such numbers as e and π in computation. Witness, again, how that supreme concept of mathematics, the
concept of *function*, has struggled through thousands of
years to win at length its present precision of form out
of the nebulous sense, which all minds have, of the mere
dependence of things on other things. Witness, too, the
mathematical concept of *infinity*, which prior to a half-
century ago was still too vague for logical discourse,
though from remotest antiquity the great idea has played
a conspicuous rôle, mainly emotional, in theology, philosophy, and science. Like examples abound, showing that
one of the most impressive and significant phenomena in
the life of the psychic world, if we will but discern and
contemplate it, is the process by which ideas advance,
often slowly indeed but surely, from their initial condition
of formlessness and indetermination to the mathematical
estate. The chemicization of biology, the physicization
of chemistry, the mechanicization of physics, the mathematicization of mechanics, the arithmeticization of mathematics, these well-known tendencies and drifts in science
do but illustrate on a large scale the ubiquitous process
in question.

At length, ladies and gentlemen, our speaker will say,
in the light of the last consideration the deeper and larger
aspects of our subject are beginning to show themselves
and there is dawning upon us an impressive vision. The
nature, function, and life of the entire conceptual world

seem to come within the circle and scope of our present enterprise. We are beginning to see that to challenge the human worth of mathematics, to challenge the worth of rigorous thinking, is to challenge the worth of all thinking, for now we see that mathematics is but the ideal to which all thinking, by an inevitable process and law of the human spirit, constantly aspires. We see that to challenge the worth of that ideal is to arraign before the bar of values what seems the deepest process and inmost law of the universe of thought. Indeed we see that in defending mathematics we are really defending a cause yet more momentous, the whole cause, namely, of the conceptual procedure of science and the conceptual activity of the human mind, for mathematics is nothing but such conceptual procedure and activity come to its maturity, purity, and perfection.

Now, ladies and gentlemen, our lecturer will say, I cannot in this course deal explicitly and fully with this larger issue. But, he will say, we are living in a day when that issue has been raised; we happen to be living in a time when, under the brilliant and effective leadership of such thinkers as Professor Bergson and the late Professor James, the method of concepts, the method of intellect, the method of science, is being powerfully assailed; and, whilst I heartily welcome this attack of criticism as causing scientific men to reflect more deeply upon the method of science, as exhibiting more clearly the inherent limitations of its method, and as showing that life is so rich as to have many precious interests and the world much truth beyond the reach of that method, yet I cannot refrain from attempting to point out what seems to me a radical error of the critics, a fundamental error of theirs, in respect to what is the highest function of conception and in respect to what is the real aim and ideal of the life of intellect. For we shall thus be led to a deeper view of our subject proper.

These critics find, as all of us find, that what we call mind or our minds is, in some mysterious way, functionally connected with certain living organisms known as human bodies; they find that these living bodies are constantly immersed in a universe of matter and motion in which they are continually pushed and pulled, heated and cooled, buffeted and jostled about — a universe that, according to James, would, in the " absence of concepts," reveal itself as " a big blooming buzzing confusion " — though it is hard to see how such a revelation could happen to any one devoid of the concept " confusion," but let that pass; our critics find that our minds get into some initial sort of knowing connection with that external blooming confusion through what they call the sensibility of our bodies, yielding all manner of sensations as of weights, pressures, pushes and pulls, of intensities and extensities of brightness, sound, time, colors, space, odors, tastes, and so on; they find that we must, on pain of organic extinction, take some account of these elements of the material world; they find that, as a fact, we human beings constantly deal with these elements through the instrumentality of concepts; they find that the effectiveness of our dealing with the material world is precisely due to our dealing with it conceptually; they infer that, therefore, dealing with matter is exactly what concepts are for, saying with Ostwald, for example, that the goal of natural science, the goal of the conceptual method of mind, " is the domination of nature by man "; not only do our critics find that we deal with the material world conceptually, and effectively because conceptually, but they find also that life has interests and the world values not accessible to the conceptual method, and as this method is the method of the intellect, they conclude, not only that the intellect cannot grasp life, but that the aim and ideal of intellect is the understanding and subjugation of matter, saying with Professor Bergson " that our intellect is intended to think matter," " that our concepts have been formed on the model of solids," " that the essential function of our intellect . . . is to be a light for our conduct, to make ready for our action on things," that " the intellect always behaves as if it were fascinated by the contemplation of inert matter," that " intelligence . . . aims at a practically useful end," that " the intellect is never quite at its ease, . . . except when it is working upon inert matter, more particularly upon solids," and much more to the same effect.

Now, ladies and gentlemen, our speaker will ask, what are we to think of this? What are we to think of this evaluation of the science-making method of concepts? What are we to think of the aim and ideal here ascribed to the intellect and of the station assigned it among the faculties of the human mind? In the first place, it ought to be evident to the critics themselves, and evident to them even in what they esteem the poor light of intellect, that the above-sketched movement of their minds is a logically unsound movement. They do not indeed contend that, because a living being in order to live must deal with the material world, it must, therefore, do so by means of concepts. The animals have taught them better. But neither does it follow that, because certain bipeds in dealing with the material world deal with it conceptually, the essential function of concepts is just to deal with matter. Nor does such an inference respecting the essential function of concepts follow from the fact that the superior effectiveness of man's dealing with the physical world is due to his dealing with it conceptually. For it is obviously conceivable and supposable that such conceptual dealing with matter is only an incident or byplay or subordinate interest in the career of concepts. It is conceivably possible that such employment with matter is only an avocation, more or less serious indeed and more or less advantageous, yet an avocation, and not the vocation, of intellect. Is it not evidently possible to go even further? Is it not logically possible to admit or to contend that, inasmuch as the human intellect is functionally attached to a living body which is itself plunged in a physical universe, it is absolutely necessary for the intellect to concern itself with matter in order to preserve, not indeed the animal life of man, but his intellectual life — is it not allowable, he will say, to admit or to maintain *that* and at the same time to deny that such concernment with matter is the intellect's chief or essential function and that the subjugation of matter is its ideal and aim?

Of course, our lecturer will say, our critics might be wrong in their logic and right in their opinion, just as they might be wrong in their opinion and right in their logic, for opinion is often a matter, not of logic or proof, but of temperament, taste, and insight. But, he will say, if the issue as to the chief function of concepts and the ideal of the intellect is to be decided in accordance with temperament, taste, and insight, then there is room for exercise of the preferential faculty, and alternatives far superior to the choice of our critics are easy enough to find. It may accord better with our insight and taste to agree with Aristotle that " It is owing," not to the necessity of maintaining animal life or the desire of subjugating matter, but " it is owing to their *wonder* that men both now begin and first began to philosophize; they *wondered* originally at the obvious difficulties, then advanced little by little and stated the difficulties about the greater matters." The striking contrast of this with the deliverances of Bergson is not surprising, for Aristotle was a pupil of Plato and the doctrine of Bergson is that of Plato completely inverted. It may accord better with our insight and taste to agree with the great K. G. J. Jacobi, who, when he had been reproached by Fourier for not devoting his splendid genius to physical investigations instead of pure mathematics, replied that a philosopher like his critic " ought to know that the unique end of science is," not public utility and application to natural phenomena, but " is the honor of the human spirit." It may accord better with our temperament and insight to agree with the sentiment of Diotima: " I am persuaded that all men do all things, and the better they are, the better they do them, in the hope," not of subjugating matter, but " in the hope of the glorious fame of immortal virtue."

But it is unnecessary, ladies and gentlemen, it is unnecessary, our speaker will say, to bring the issue to final trial in the court of temperaments and tastes. We should gain there a too easy victory. The critics are psychologists, some of them eminent psychologists. Let the issue be tried in the court of psychology, for it is there that of right it belongs. They know the fundamental and relevant facts. What is the verdict according to these? The critics know the experiments that have led to and confirmed the psychophysical law of Weber and Fechner and the doctrine of thresholds; they know that, in accordance with that doctrine and that law, an appropriate stimulus, no matter what the department of sense, may be finite in amount and yet too small, or finite and yet too large, to yield a sensation; they know that the difference between two stimuli of a kind appropriate to a given sense department, no matter what department, may be a finite difference and yet too small for sensibility to detect, or to work a change of sensation; they ought to know, though they seem not to have recognized, much less to have weighed, the fact that, owing to the presence of thresholds, the greatest number of distinct sensations possible in any department of sense is a *finite* number; they ought to know that the number of different departments of sense is also a *finite* number; they ought to know that, therefore, the total number of distinct or different *sensations* of which a human being is capable is a *finite* number; they ought to know, though they seem not to have recognized the fact, that, on the other hand, the world of *concepts* is of *infinite* multiplicity, that *concepts,* the fruit of intellect, as distinguished from sensations, the fruit of sensibility, are *infinite* in number; they ought, therefore, to see, our speaker will say, though none of them has seen, that in attempting to derive intellect out of sensibility, in attempting to show that (as James says) " concepts flow out of percepts," they are confronted with the problem of bridging the immeasurable gulf between the finite and the infinite, of showing, that is, how an infinite multiplicity can arise from one that is finite. But even if they solved that apparently insuperable problem, they could not yet be in position to affirm that the function of intellect and its concepts is, like that of sensibility, just the function of dealing with matter, as the function of teeth is biting and chewing. Far from it.

Let us have another look, the lecturer will say, at the psychological facts of the case. Owing to the presence of thresholds in every department of sense it may happen and indeed it does happen constantly in every department, that three different amounts of stimulus of a same kind give *three sensations such that two of them are each indistinguishable from the third and yet are distinguishable from one another.* Now, for sensibility in any department of sense, two magnitudes of stimulus are unequal or equal according as the sensations given by them are or are not distinguishable. Accordingly in the world of sensible magnitudes, in the sensible universe, in the world, that is, of *felt* weights and thrusts and pulls and pressures, of *felt* brightnesses and warmths and lengths and breadths and thicknesses and so on, in this world, which is the world of matter, *magnitudes are such that two of them may each be equal to a third without being equal to one another.* That, our speaker will say, is a most significant fact and it means that the sensible world, the world of matter, is irrational, infected with contradiction, contravening the essential laws of thought. No wonder, he will say, that old Heracleitus declared the unaided senses " give a fraud and a lie."

Now, our speaker will ask, what has been and is the behavior of intellect in the presence of such contradiction? Observe, he will say, that it is intellect, and not sensibility, that detects the contradiction. Of the irrationality in question sensibility remains insensible. The data among which the contradiction subsists are indeed rooted in the sensible world, they inhere in the world of matter, but the contradiction itself is known only to the logical faculty called intellect. Observe also, he will say, and the observation is important, that such contradictions do not compel the intellect to any activity whatever intended to preserve the life of the living organism to which the intellect is functionally attached. That is a lesson we have from our physical kin, the beasts. What, then, *has* the intellect done because of or about the contradiction? Has it gone on all these centuries, as our critics would have us believe, trying to " think matter," as if it did not know that matter, being irrational, is not thinkable? Far from it, he will say, the intellect is no such ass.

What it has done, instead of endlessly and stupidly besieging the illogical world of sensible magnitudes with the machinery of logic, what it has done, our lecturer will say, is this: it has created for itself another world. It has not rationalized the world of sensible magnitudes. That, it knows, cannot be done. It has discerned the ineradicable contradictions inherent in them, and by means of its creative power of conception it has made a new world, a world of conceptual magnitudes that, like the *continua* of mathematics, are so constructed by the spiritual architect and so endowed by it as to be free alike from the contradictions of the sensible world and from all thresholds that could give them birth. Indeed conception, to speak metaphorically in terms borrowed from the realm of sense, is a kind of infinite sensibility, transcending any finite distinction, difference or threshold, however minute or fine. And now, our speaker will say, it is such magnitudes, magnitudes created by intellect and not those discovered by sense, though the two varieties are frequently not discriminated by their names; it is such conceptual magnitudes that constitute the subject-matter of science. If the magnitudes of science, apart from their rationality, often bear in conformation a kind of close resemblance to magnitudes of sense, what is the meaning of the fact? It means, contrary to the view of Bergson but in accord with that of Poincaré, that the free creative artist, intellect, though it is not constrained, yet has chosen to be guided, in so far as its task allows, by facts of sense. Thus we have, for example, conceptual space and sensible space so much alike in conformation that, though one of them is rational and the other is not, the undiscriminating hold them as the same.

And now, our lecturer will ask, for we are nearing the goal, what, then, is the motive and aim of this creative activity of the intellect? Evidently it is not to preserve and promote the life of the human body, for animals flourish without the aid of concepts, without " discourse of reason," and despite the contradictions in the world of sense. The aim is, he will say, to preserve and promote the life of the intellect itself. In a realm infected with irrationality, with omnipresent contradictions of the laws of thought, intellect cannot live, much less flourish; in the world of sense, it has no proper subject-matter, no home, no life. To live, to flourish, it must be able to *think*, to think in accordance with the laws of its being. It is stimulated and its activity is sustained by two opposite forces: discord and concord. By the one it is driven; by the other, drawn. Intellect is a perpetual suitor. The object of the suit is, not the conquest of matter, it is a thing of mind, it is the music of the spirit, it is Harmonia, the beautiful daughter of the Muses. The aim, the ideal, the beatitude of intellect is harmony. That is the meaning of its endless talk about compatibilities, consistencies and concords, and that is the meaning of its endless battling and circumvention and transcendence of contradiction. But what of the applications of science and public service? These are by-products of the intellect's aim and of the pursuit of its ideal. Many things it regards as worthy, high, and holy — applications of science, public service, the " wonder " of Aristotle, Jacobi's " honor of the human spirit," Diotima's " glorious fame of immortal virtue " — but that which, by the law of its being, Intellect seeks above all and perpetually pursues and loves, is Harmony. It is for a home and a dwelling with her that intellect creates a world; and its admonition is: Seek ye first the kingdom of harmony, and all these things shall be added unto you.

And the ideal and admonition, thus revealed in the light of analysis, are justified of history. Inverting the order of time, we have only to contemplate the great periods in the intellectual life of Paris, Florence, and Athens. If, among these mightiest contributors to the spiritual wealth of man, Athens is supreme, she is also supreme in her devotion to the intellect's ideal. It is of Athens that Euripides sings:

> The sons of Erechtheus, the olden,
>   Whom high gods planted of yore
> In an old land of heaven upholden,
>   A proud land untrodden of war:
> They are hungered, and lo, their desire
>   With wisdom is fed as with meat:
> In their skies is a shining of fire,
>   A joy in the fall of their feet:
> And thither with manifold dowers,
>   From the North, from the hills, from the morn,
> The Muses did gather their powers,
>   That a child of the Nine should be born;
> And Harmony, sown as the flowers,
>   Grew gold in the acres of corn.[4]

And thus, ladies and gentlemen, our lecturer will say, what I wish you to see here is, that science and especially mathematics, the ideal form of science, are creations of the intellect in its quest of harmony. It is as such creations that they are to be judged and their human worth appraised. Of the applications of mathematics to engineering and its service in natural science, I have spoken at length, he will say, in course of previous lectures. Other great themes of our subject remain for consideration. To appraise the worth of mathematics as a

[4] Translation by Professor Gilbert Murray.

discipline in the art of rigorous thinking and as a means of giving facility and wing to the subtler imagination; to estimate and explain its value as a norm for criticism and for the guidance of speculation and pioneering in fields not yet brought under the dominion of logic; to estimate its esthetic worth as showing forth in psychic light the law and order of the psychic world; to evaluate its ethical significance in rebuking by its certitude and eternality the facile scepticism that doubts all knowledge, and especially in serving as a retreat for the spirit when as at times the world of sense seems madly bent on heaping strange misfortunes up and " to and fro the chances of the years dance like an idiot in the wind "; to give a sense of its religious value in " the contemplation of ideas under the form of eternity," in disclosing a cosmos of perfect beauty and everlasting order and in presenting there, for meditation, endless sequences traversing the rational world and seeming to point to a mystical region above and beyond; [5] these and similar themes, our speaker will say, remain to be dealt with in subsequent lectures of the course.

[5] For a development of this notion of an Over-world see Keyser's *Science and Religion,* The Yale University Press.

THE HUMAN SIGNIFICANCE OF MATHEMATICS [1]

Homo sum; humani nil a me alienum puto.
— TERENCE

THE subject of this address is not of my choosing. It came to me by assignment. I may, therefore, be allowed to say that it is in my judgment ideally suited to the occasion. This meeting is held here upon this beautiful coast because of the presence of an international exposition, and we are thus invited to a befitting largeness and liberality of spirit. An international exposition properly may and necessarily will admit many things of a character too technical to be intelligible to any one but the expert and the specialist. Such things, however, are only incidental — contributory, indeed, yet incidental — to pursuit of the principal aim, which is, I believe, or ought to be, the representation of human things as human — an exhibition and interpretation of industries, institutions, sciences and arts, not primarily in their accidental or particular character as illustrating individuals or classes or specific localities or times, but primarily in their essential and universal character as representative of man. A world-exposition will, therefore, as far as practicable, avoid placing in the forefront matters so abstruse as to be fit for the contemplation and understanding of none

[1] An address delivered August 3, 1915, Berkeley, Calif., at a joint meeting of the American Mathematical Society, the American Astronomical Society, and Section A of the American Association for the Advancement of Science. Printed in *Science*, November 12, 1915.

but specialists; it will, as a whole, and in all its principal parts, address itself to the general intelligence; for it aims at being, for the multitudes of men and women who avail themselves of its exhibitions and lessons, an exposition of humanity: an exposition, no doubt, of the activities and aspirations and prowess of individual men and women, but of men and women, not in their capacity as individuals, but as representatives of humankind. Individual achievements are not the object, they are the means, of the exposition. The object is humanity.

What is the human significance — what is the significance for humanity — of "the mother of the sciences"? And how may the matter be best set forth, not for the special advantage of professional mathematicians, for I shall take the liberty of having these but little in mind, but for the advantage and understanding of educated men and women in general? I am unable to imagine a more difficult undertaking, so technical, especially in its language, and so immense is the subject. It is clear that the task is far beyond the resources of an hour's discourse, and so it is necessary to restrict and select. This being the case, what is it best to choose? The material is superabundant. What part of it or aspect of it is most available for the end in view? "In abundant matter to speak a little with elegance," says Pindar, "is a thing for the wise to listen to." It is not, however, a question of elegance. It is a question of emphasis, of clarity, of effectiveness. What shall be our major theme?

Shall it be the history of the subject? Shall it be the modern developments of mathematics, its present status and its future outlook? Shall it be the utilities of the science, its so-called applications, its service in practical affairs, in engineering and in what it is customary to call

the sciences of nature? Shall it be the logical foundations of mathematics, its basic principles, its inner nature,
its characteristic processes and structure, the differences
and similitudes that come to light in comparing it with
other forms of scientific and philosophic activity? Shall
it be the bearings of the science as distinguished from its
applications — the bearings of it as a spiritual enterprise
upon the higher concerns of man as man? It might be
any one of these things. They are all of them great and
inspiring themes.

It is easy to understand that a historian would choose
the first. The history of mathematics is indeed impressive, but is it not too long and too technical? And is it
not already accessible in a large published literature of
its own? I grant, the historian would say, that its
history is long, for in respect of antiquity mathematics
is a rival of art, surpassing nearly all branches of science and by none of them surpassed. I grant that, for
laymen, the history is technical, frightfully technical, requiring interpretation in the interest of general intelligence. I grant, too, that the history owns a large literature, but this, the historian would say, is not designed
for the general reader, however intelligent, the numerous
minor works no less than the major ones, including that
culminating monumental work of Moritz Cantor, being,
all of them, addressed to specialists and intelligible to
them alone. And yet it would be possible to tell in one
hour, not indeed the history of mathematics, but a true
story of it that would be intelligible to all and would show
its human significance to be profound, manifold, and even
romantic. It would be possible to show historically that
this science, which now carries its head so high in the
tenuous atmosphere of pure abstractions, has always kept
its feet upon the solid earth; it would be possible to show

that it owns indeed a lowly origin, in the familiar needs
of common life, in the homely necessities of counting
herds and measuring lands; it would be possible to show
that, notwithstanding its birth in the concrete things of
sense and raw reality, it yet so appealed to sheer intellect
— and we must not forget that creative intellect is the
human faculty par excellence — it so appealed to this
distinctive and disinterested faculty of man that, long
before the science rose to the level of a fine art in
the great days of Euclid and Archimedes, Plato in the
wisdom of his maturer years judged it essential to the
education of freemen because, said he, there is in it a
necessary something against which even God can not
contend and without which neither gods nor demi-gods
can wisely govern mankind; it would be possible, our
historian could say, to show historically to educated
laymen that, even prior to the inventions of analytical
geometry and infinitesimal calculus, mathematics had
played an indispensable rôle in the " Two New Sciences "
of physics and mechanics in which Galileo laid the
foundations of our modern knowledge of nature; it
would be possible to show not only that the analytical
geometry of Descartes and Fermat and the calculus
of Leibniz and Newton have been and are essential to
our still advancing conquest of the sea, but that it is
owing to the power of these instruments that the genius
of such as Newton, Laplace and Lagrange has been enabled to create for us a new earth and a new heavens
compared with which the Mosaic cosmogony or the sublimest creation of the Greek imagination is but " as a
cabinet of brilliants, or rather a little jewelled cup found
in the ocean or the wilderness "; [2] it would be possible
to show historically that, just because the pursuit of

[2] James Martineau: *The Seat of Authority in Religion.*

mathematical truth has been for the most part disinterested — led, that is, by wonder, as Aristotle says, and sustained by the love of beauty with the joy of discovery — it would be possible to show that, just because of the disinterestedness of mathematical research, this science has been so well prepared to meet everywhere and always, as they have arisen, the mathematical exigencies of natural science and engineering; above all, it would be feasible to show historically that to the same disinterestedness of motive operating through the centuries we owe the upbuilding of a body of pure doctrine so towering to-day and vast that no man, even though he have the " Andean intellect " [3] of a Poincaré, can embrace it all. This much, I believe, and perhaps more, touching the human significance of mathematics, a historian of the science might reasonably hope to demonstrate in one hour.[4]

More difficult, far more difficult, I think, would be the task of a pure mathematician who aimed at an equivalent result by expounding, or rather by delineating, for he could not in one hour so much as begin to expound, the modern developments of the subject. Could he contrive even to delineate them in a way to reveal their relation to what is essentially humane? Do but consider for a moment the nature of such an enterprise. Mathematics may be legitimately pursued for its own sake or for the sake of its applications or with a view to understanding its logical foundations and internal structure or in the interest of magnanimity or for the sake of its bearings upon the supreme concerns of man as man or from two or more of these motives combined. Our supposed

[3] The phrase is due to Professor Wm. Benj. Smith.
[4] It is a pleasure to refer the reader to Professor David Eugene Smith's *History of Mathematics,* Ginn and Company.

delineator is actuated by the first of them: his interest in
mathematics is an interest in mathematics for the sake of
mathematics; for him the science is simply a large and
growing body of logical consistencies or compatibilities;
he derives his inspiration from the muse of intellectual
harmony; he is a pure mathematician. He knows that
pure mathematics is a house of many chambers; he
knows that its foundations lie far beneath the level of
common thought; and that the superstructure, quickly
transcending the power of imagination to follow it, ascends higher and higher, ever keeping open to the sky;
he knows that the manifold chambers — each of them a
mansion in itself — are all of them connected in wondrous ways, together constituting a fit laboratory and
dwelling for the spirit of men of genius. He has assumed
the task of presenting a vision of it that shall be worthy
of a world-exposition. Can he keep the obligation? He
wishes to show that the life and work of pure mathematicians are human life and work: he desires to show
that these toilers and dwellers in the chambers of pure
thought are representative men. He would exhibit the
many-chambered house to the thronging multitudes of
his fellow men and women; he would lead them into it;
he would conduct them from chamber to chamber by the
curiously winding corridors, passing now downward, now
upward, by delicate passageways and subtle stairs; he
would show them that the wondrous castle is not a dead
or static affair like a structure of marble or steel, but a
living architecture, a living mansion of life, human as
their own; he would show them the mathetic spirit at
work, how it is ever weaving, tirelessly weaving, fabrics
of beauty, finer than gossamer yet stronger than cables
of steel; he would show them how it is ever enlarging its
habitation, deepening its foundations, expanding more

and more and elevating the superstructure; and, what is even more amazing, how it perpetually performs the curious miracle of permanence combined with change, transforming, that is, the older portions of the edifice without destroying it, for the structure is eternal: in a word, he would show them a vision of the whole, and he would do it in a way to make them perceive and feel that, in thus beholding there a partial and progressive attainment of the higher ideals of man, they were but gazing upon a partial and progressive realization of their own appetitions and dreams.

That is what he *would* do. But how? *Mengenlehre, Zahlenlehre,* algebras of many kinds, countless geometries of countless infinite spaces, function theories, transformations, invariants, groups and the rest — how can these with all their structural finesse, with their heights and depths and limitless ramifications, with their labyrinthine and interlocking modern developments — I will not say how can they be presented in the measure and scale of a great exposition — but how is it possible in one hour to give laymen even a glimpse of the endless array? Nothing could be more extravagant or more absurd than such an undertaking. Compared with it, the American traveler's hope of being able to see Rome in a single forenoon was a most reasonable expectation. But it is worth while trying to realize how stupendous the absurdity is.

It is evident that our would-be delineator must compromise. He can not expound, he can not exhibit, he can not even delineate the doctrines whose human worth he would thus disclose to his fellow men and women. The fault is neither his nor theirs. It must be imputed to the nature of things. But he need not, therefore, despair and he need not surrender. The method he has proposed — the method of exposition — that indeed he must abandon as hopeless, but not his aim. He is addressing men and women who are no doubt without his special knowledge and his special discipline, as he in his turn is without theirs, but who are yet essentially like himself. He would have them as fellows and comrades persuaded of the dignity of his *Fach:* he would have them feel that it is also theirs; he would have them convinced that mathematics stands for an immense body of human achievements, for a diversified continent of pure doctrine, for a discovered world of intellectual harmonies. He can not *show* it to them as a painter displays a canvas or as an architect presents a cathedral. He can not give them an immediate vision of it, but he can give them intimations; by appealing to their *fantasie* and, through analogy with what they know, to their understanding, not only can he convince them that his world exists, but he can give them an intuitive apprehension of its living presence and its meaning for humankind. This is possible because, like him, they, too, are idealists, dreamers and poets — such essentially are all men and women. His auditors or his readers have all had *some* experience of ideas and of truth, they have all had inklings of more beyond, they have all been visited and quickened by a sense of the limitless possibilities of further knowledge in every direction, they have all dreamed of the perfect and have felt its lure. They are thus aware that the small implies the large; having seen hills, they can believe in mountains; they know that Euripides, Shakespeare, Dante, Goethe, are but fulfillments of prophecies heard in peasant tales and songs; they know that the symphonies of Beethoven or the dramas of Wagner are harbingered in the melodies and the sighs of those who garner grain and in their hearts respond to the music of the winds or the " solemn anthems of the sea " ; they

sense the secret by which the astronomy of Newton and
Laplace is foretokened in the shepherd's watching of the
stars; and knowing thus this plain spiritual law of progressiveness and implication, they are prepared to grasp
the truth that modern mathematics, though they do not
understand it, is, like the other great things, but a sublime fulfillment, the realization of prophecies involved in
what they themselves, in common with other educated
folk, know of the rudiments of the science. Indeed, they
would marvel if upon reflection it did not seem to be so.
Our pure mathematician in speaking to his fellow men
and women of his science will have no difficulty in per-
suading them that he is speaking of a subject immense
and eternal. As born idealists they have intimations of
their own — the evidence of intuition, if you please — or
a kind of insight resembling that of the mystic — that in
the world of mind there must be something deeper and
higher, stabler and more significant, than the pitiful ideas
in life's routine and the familiar vocations of men. They
are thus prepared to believe, before they are told, that
behind the veil there exists a universum of exact thought,
an everlasting cosmos of ordered ideas, a stable world of
concatenated truth. In their study of the elements, in
school or college, they may have caught a shimmer of it
or, in rare moments of illumination, even a gleam. Of
the existence, the reality, the actuality, of our pure
mathematician's world they will have no doubt, and they
will have no doubt of its grandeur. They may even, in
a vague way, magnify it overmuch, feeling that it is, in
some wise, *more* than human, significant only for the
rarely gifted spirit that dwells, like a star, apart. The
pure mathematician's difficulty lies in showing, in *his* way,
that such is not the case. For he does not wish to adduce
utilities and applications. He is well aware of these. He

knows that if he " would tell them they are more in number than the sands." Neither does he despise them as of little moment. On the contrary, he values them as precious. But he wishes to do his subject and his auditors the honor of speaking from a higher level: he desires to vindicate the worth of mathematics on the ground of its sheer ideality, on the ground of its intellectual harmony, on the ground of its beauty, " free from the gorgeous trappings " of sense, pure, austere, supreme. To do this, which ought, it seems, to be easy, experience has shown to be exceedingly difficult. For the multitude of men and women, even the educated multitude, are wont to cry,

> Such knowledge is too wonderful for me,
> It is too high, I can not attain unto it,

thus meaning to imply, What, then, or where is its human significance? Their voice is heard in the challenge once put to me by the brilliant author [5] of " East London Visions." What, said he, can be the human significance of " this majestic intellectual cosmos of yours, towering up like a million-lustred iceberg into the arctic night," seeing that, among mankind, none is permitted to behold its more resplendent wonders save the mathematician alone? What response will our pure mathematician make to this challenge? Make, I mean, if he be not a wholly naïve devotee of his science and so have failed to reflect upon the deeper grounds of its justification. He may say, for one thing, what Professor Klein said on a similar occasion:

Apart from the fact that pure mathematics can not be supplanted by anything else as a means for developing the purely logical faculties of the mind, there must be considered here, as elsewhere, the necessity of the presence of a few individuals in each country developed in far

[5] Edward Willmore.

higher degree than the rest, for the purpose of keeping up and gradually raising the *general* standard. Even a slight raising of the general level can be accomplished only when some few minds have progressed far ahead of the average.

That is doubtless a weighty consideration. But is it all or the best that may be said? It is just and important but it does not go far enough; it is not, I fear, very convincing; it is wanting in pungence and edge; it does not touch the central nerve of the challenge. Our pure mathematician must rally his sceptics with sharper considerations. He may say to them: You challenge the human significance of the higher developments of pure mathematics because they are inaccessible to all but a few, because their charm is esoteric, because their deeper beauty is hid from nearly all mankind. Does that consideration justify your challenge? You are individuals, but˙you are also members of a race. Have you as individuals no human interest nor human pride in the highest achievements of your race? Is nothing human, is nothing humane, except mediocrity and the commonplace? Was Phidias or Michel Angelo less human than the carver and painter of a totem-pole? Was Euclid or Gauss or Poincaré less representative of man than the countless millions for whom mathematics has meant only the arithmetic of the market place or the rude geometry of the carpenter? Does the quality of humanity in human thoughts and deeds decrease as they ascend towards the peaks of achievement, and increase in proportion as they become vulgar, attaining an upper limit in the beasts? Do you not know that precisely the reverse is true? Do you not count aspiration humane? Do you not see that it is not the common things that every one may reach, but excellences high-dwelling among the rocks — do you not know that, in respect of human worth, these things, which but

few can attain, are second only to the supreme ideals
attainable by none?

How very different and how very much easier the task
of one who sought to vindicate the human significance
of mathematics on the ground of its applications! In
respect of temperamental interest, of attitude and outlook, the difference between the pure and the applied
mathematician is profound. It is — if we may liken spiritual things to things of sense — much like the difference
between one who greets a new-born day because of its
glory and one who regards it as a time for doing chores
and values its light only as showing the way. For the
former, mathematics is justified by its supreme beauty;
for the latter, by its manifold use. But are the two kinds
of value essentially incompatible? They are certainly
not. The difference is essentially a difference of authority
— a difference, that is, of worth, of elevation, of excellence. The pure mathematician and the applied mathematician sometimes may, indeed they not infrequently do,
dwell together harmoniously in a single personality. If
our spokesman be such a one — and I will not suppose
the shame of having the utilities of the science represented on such an occasion by one incapable of regarding
it as anything but a tool, for that would be disgraceful —
if, then, our spokesman be such a one as I have supposed,
he might properly begin as follows: In speaking to you
of the applications of mathematics I would not have you
suppose, ladies and gentlemen, that I am thus presenting
the *highest* claims of the science to your regard; for its
highest justification is the charm of its immanent beauty;
I do not mean, he will say, the beauty of appearances —
the fleeting beauties of sense, though these, too, are
precious — even the outer garment, the changeful robe,
of reality is a lovely thing; I mean the eternal beauty of

the world of pure thought; I mean intellectual beauty; in mathematics this nearly attains perfection; and " intellectual beauty is self-sufficing " ; uses, on the other hand, are not; they wear an aspect of apology; uses resemble excuses, they savor a little of a plea in mitigation. Do you ask: Why, then, plead them?   Because, he will say, many good people have a natural incapacity to appreciate anything else; because, also, many of the applications, especially the higher ones, are themselves matters of exceeding beauty; and especially because I wish to show, not only that use and beauty are compatible forms of worth, but that the more mathematics has been cultivated for the sake of its inner charm, the fitter has it become for external service.

Having thus at the outset put himself in proper light and given his auditors a scholar's warning against what would else, he fears, foster a disproportionment of values, what will he go on to signalize among the utilities of a science whose primary allegiance to logical rectitude allies it to art, and which only incidentally and secondarily shapes itself to the ends of instrumental service?   He knows that the applications of mathematics, if one will but trace them out in their multifarious ramifications, are as many-sided as the industries and as manifold as the sciences of men, penetrating everywhere throughout the full round of life.   What will he select?   He will not dwell long upon its homely uses in the rude computations and mensurations of counting-house and shop and factory and field, for this indispensable yet humble manner of world-wide and perpetual service is known of all men and women.   He will quickly pass to higher considerations — to navigation, to the designing of ships, to the surveying of lands and seas, and the charting of the world, to the construction of reservoirs and aqueducts, canals, tunnels and railroads, to the modern miracles of the marine cable, the telegraph, the telephone, to the multiform achievements of every manner of modern engineering, civil, mechanical, mining, electrical, by which, through the advancing conquest of land and sea and air and space and time, the conveniences and the prowess of man have been multiplied a billionfold. It need not be said that not all this has been done by mathematics alone. Far from it. It is, of course, the joint achievement of many sciences and arts, but — and just this is the point — the contributions of mathematics to the great work, direct and indirect, have been indispensable. And it will require no great skill in our speaker to show to his audience, if it have a little imagination, that, as I have said elsewhere, if all these mathematical contributions were by some strange spiritual cataclysm to be suddenly withdrawn, the life and body of industry and commerce would suddenly collapse as by a paralytic stroke, the now splendid outer tokens of material civilization would quickly perish, and the face of our planet would at once assume the aspect of a ruined and bankrupt world. For such is the amazing utility, such the wealth of by-products, if you please, that come from a science and art that owes its life, its continuity and its power to man's love of intellectual harmony and pleads its inner charm as its sole appropriate justification. Indeed it appears — contrary to popular belief — that in our world there is nothing else quite so practical as the inspiration of a muse.

But this is not all nor nearly all to which our applied mathematician will wish to invite attention. It is only the beginning of it. Even if he does not allude to the quiet service continuously and everywhere rendered by mathematics in its rôle as a norm or standard or ideal in every field of thought whether exact or inexact, he will yet desire to instance forms and modes of application compared with which those we have mentioned, splendid and impressive as they are, are meagre and mean. For those we have mentioned are but the more obvious applications — those, namely, that continually announce themselves to our senses everywhere in the affairs, both great and small, of the workaday world. But the really great applications of mathematics — those which, rightly understood, best of all demonstrate the human significance of the science — are not thus obvious; they do not, like the others, proclaim themselves in the form of visible facilities and visible expedients everywhere in the offices, the shops, and the highways of commerce and industry; they are, on the contrary, almost as abstract and esoteric as mathematics itself, for they are the uses and applications of this science in other sciences, especially in astronomy, in mechanics and in physics, but also and increasingly in the newer sciences of chemistry, geology, mineralogy, botany, zoology, economics, statistics and even psychology, not to mention the great science and art of architecture. In the matter of exhibiting the endless and intricate applications of mathematics to the natural sciences, applications ranging from the plainest facts of crystallography to the faint bearings of the kinetic theory of gases upon the constitution of the Milky Way, our speaker's task is quite as hopeless as we found the *pure* mathematician's to be; and he, too, will have to compromise; he will have to request his auditors to acquaint themselves at their leisure with the available literature of the subject and especially to read attentively the great work of John Theodore Merz dealing with the " History of European Thought in the Nineteenth Century," where they will find, in a form fit for the general reader, how central has been the rôle of mathematics in all the principal attempts of natural science to find a cosmos in the seeming chaos of the natural world. Another many-sided work that in this connection he may wish to commend as being in large part intelligible to men and women of general education and catholic mind is Enriques's " Problems of Science."

I turn now for a moment to the prospect of one who might choose to devote the hour to an exposition or an indication of modern developments in what it is customary to call the foundations of mathematics — to a characterization, that is, and estimate of that far-reaching and still advancing critical movement which has to do with the relations of the science, philosophically considered, to the sciences of logic and methodology. What can he say on this great theme that will be intelligible and edifying to the multitudes of men and women who, though mathematically inexpert, yet have a genuine humane curiosity respecting even the profounder and subtler life and achievements of science? He can point out that mathematics, like all the other sciences, like the arts too, for that matter, and like philosophy, originates in the refining process of reflection upon the crude data of common sense; he can point out that this process has gradually yielded from out the raw material and still continues to yield more and more ideas of approximate perfection in the respects of precision and form; he can point out that such ideas, thus disentangled and trimmed of their native vagueness and indetermination, disclose their mutual relationships and so become amenable to the concatenative processes of logic; and he can point out that these polished ideas with their mutual relationships become the bases or the content of various branches of mathematics, which thus tower above common sense and appear to grow out of it and to stand upon it like trees or forests upon the earth. He will point out, however, that this appearance, like most other obvious appearances, is deceiving; he will, that is, point out that these upward-growing sciences or branches of science are found, in the light of further reflection, to be downward-growing as well, pushing their roots deeper and deeper into a dark soil far beneath the ground of evident common sense; indeed, he will show that common sense is thus, in its relation to mathematics, but as a sense-litten mist enveloping only the mid-portion of the stately structure, which, like a towering mountain, at once ascends into the limpid ether far above the shining cloud and rests upon a base of subterranean rock far below; he will point out that, accordingly, mathematicians, in respect of temperamental interest, fall into two classes — the class of those who cultivate the upward-growing of the science, working thus in the upper regions of clearer light, and the class of those who devote themselves to exploring the deep-plunging roots of the science; and it is, he will say, to the critical activity of the latter class — the logicians and philosophers of mathematics — that we owe the discovery of what we are wont to call the foundations of mathematics — the great discovery, that is, of an immense mathematical *sub*-structure, which penetrates far beneath the stratum of common sense and of which many of even the greatest mathematicians of former times were not aware. But whilst such foundational research is in the main a modern phenomenon, it is by no means exclusively such; and to protect his auditors against a false perspective in this regard and the peril of an overweening pride in the achievements of their own time, our speaker may recommend to them the perusal of Thomas L. Heath's superb edition of Euclid's " Elements " where, especially in the first volume, they will be much edified to find, in the rich abundance of critical citation and commentary which the translator has there brought together, that the refined and elaborate logico-mathematical researches of our own time have been only a deepening and widening of the keen mathematical criticism of a few centuries immediately preceding and following the great date of Euclid. Indeed but for that general declension of Greek spirit which Professor Gilbert Murray in his " Four Stages of Greek Religion " has happily characterized as " the failure of nerve," what we know as the modern critical movement in mathematics might well have come to its present culmination, so far at least as pure geometry is concerned, fifteen hundred or more years ago. It is a pity that the deeper and stabler things of science and the profounder spirit of man can not be here disclosed in a manner commensurate with the great exposition, surrounding us, of the manifold practical arts and industries of the world. It is a pity there is no means by which our speaker might, in a manner befitting the subject and the occasion, exhibit intelligibly to his fellow men and women the ways and results of the last hundred years of research into the groundwork of mathematical science and therewith the highly important modern developments in logic and the theory of knowledge. How astonished the beholders would be, how delighted too, and proud to belong to a race capable of such patience and toil, of such disinterested devotion, of such intellectual finesse and depth of penetration. I can think of no other spectacle quite so impressive as the inner vision of all the manifold branches of rigorous thought seen to constitute one immense structure of autonomous doctrine reposing upon the spiritual basis of a few select ideas and, superior to the fading beauties of time and sense, shining there like a celestial city, in " the white radiance of eternity." [6]   That

[6] From Shelley.

is the vision of mathematics that a student of its philosophy would, were it possible, present to his fellow men and women.

In view of the foregoing considerations it evidently is, I think, in the nature of the case impossible to give an adequate sense of the human worth of mathematics if one choose to devote the hour to any one of the great aspects of it with which we have been thus far concerned. Neither the history of the subject nor its present estate nor its applications nor its logical foundations — no one of these themes lends itself well to the purpose of such exposition, and still less do two or more of them combined. Even if such were not the case I should yet feel bound to pursue another course; for I have been long persuaded that, in respect of its human significance, mathematics invites to a point of view which, unless I am mistaken, has not been taken and held in former attempts at appreciation. I have already alluded to bearings of mathematics as distinguished from applications. It is with its bearings that I wish to deal. I mean its bearings upon the higher concerns of man as man — those interests, namely, which have impelled him to seek, over and above the needs of raiment and shelter and food, some inner adjustment of life to the poignant limitations of life in our world and which have thus drawn him to manifold forms of wisdom, not only to mathematics and natural science, but also to literature and philosophy, to religion and art, and theories of righteousness. What is the rôle of mathematics in this perpetual endeavor of the human spirit everywhere to win reconciliation of its dreams and aspirations with the baffling conditions and tragic facts of life and the world? What is its relation to the universal quest of man for some supreme and abiding good that shall assuage or annul the discords and tyrannies of time and limitation, withholding less and less, as time goes by, the freedom and the peace of an ideal harmony infinite and eternal?

In endeavoring to suggest, in the time remaining for this address, a partial answer to that great question, in attempting, that is, to indicate the relations of mathematics to the supreme ideals of mankind, it will be necessary to seek a perspective point of view and to deal with large matters in a large way.

Of the countless variety of appetitions and aspirations that have given direction and aim to the energies of men and that, together with the constraining conditions of life in our world, have shaped the course and determined the issues of human history, it is doubtless not yet possible to attempt confident and thoroughgoing classification according to the principle of relative dignity or that of relative strength. If, however, we ask whether, in the great throng of passional determinants of human thought and life, there is one supreme passion, one that in varying degrees of consciousness controls the rest, unifying the spiritual enterprises of our race in directing and converging them all upon a single sovereign aim, the answer, I believe, can not be doubtful: the activities and desires of mankind are indeed subject to such imperial direction and control. And if now we ask what the sovereign passion is, again the answer can hardly admit of question or doubt. In order to see even *a priori* what the answer must be, we have only to imagine a race of beings endowed with our human craving for stability, for freedom, and for perpetuity of life and its fleeting goods, we have only to fancy such a race flung, without equipment of knowledge or strength, into the depths of a treacherous universe of matter and force where they are tossed, buffeted and torn by the tumultuous onward-rushing flood of the cosmic stream, originating they know not whence and flowing they know not why nor whither, we have, I say, only to imagine *this*, sympathetically, which ought to be easy for us as men, and then to ask ourselves what would *naturally* be the controlling passion and dominant enterprise of such a race — unless, indeed, we suppose it to become strangely enamored of distress or to be driven by despair to self-extinction. We humans require no Gotama nor Heracleitus to tell us that man's lot is cast in a world where naught abides. The universal impermanence of things, the inevitableness of decay, the mocking frustration of deepest yearnings and fondest dreams, all this has been keenly realized wherever men and women have had seeing eyes or been even a little touched with the malady of meditation, and everywhere in the literature of power is heard the cry of the mournful truth. " The life of man," said the Spirit of the Ocean, " passes by like a galloping horse, changing at every turn, at every hour."

> " Great treasure halls hath Zeus in heaven,
> From whence to man strange dooms be given,
> Past hope or fear."

Such is the universal note. Whether we glance at the question in a measure *a priori,* as above, or look into the cravings of our own hearts, or survey the history of human emotion and thought, we shall find, I think, in each and all these ways, that human life owns the supremacy of one desire: it is the passion for emancipation, for release from life's limitations and the tyranny of change: it is our human passion for some ageless form of reality, some everlasting vantage-ground or rock to stand upon, some haven of refuge from the all-devouring transformations of the weltering sea. And so it is that our human aims, aspirations, and toils thus find their highest unity — their only intelligible unity — in the spirit's quest of a stable world, in its endless search for some mode or form of reality that is at once infinite, changeless, eternal.

Does some one say: This may be granted, but what is the point of it all? It is obviously true enough, but what, pray, can be its bearing upon the matter in hand? What light does it throw upon the human significance of mathematics? The question is timely and just. The answer, which will grow in fullness and clarity as we proceed, may be at once begun.

How long our human ancestors, in remote ages, may have groped, as some of their descendants even now grope, among the things of *sense,* in the hope of finding *there* the desiderated good, we do not know — past time is long and the evolution of wisdom has been slow. We do know that, long before the beginnings of recorded history, superior men — advanced representatives of their kind — must have learned that the deliverance sought was not to be found among the objects of the *mobile* world, and so the spirit's quest passed from thence; passed from the realm of perception and sense to the realm of concept and reason: thought ceased, that is, to be merely the unconscious means of pursuit and became itself the quarry — mind had discovered mind; and there, in the realm of ideas, in the realm of spirit proper, in the world of reason or thought, the great search — far outrunning historic time — has been endlessly carried on, with varying fortunes, indeed, but without despair or breach of continuity, meanwhile multiplying its resources and assuming gradually, as the years and centuries have passed, the characters and forms of what we know today as philosophy and science and art. I have mentioned the passing of the quest from the realm of sense to the realm of conception: a most notable transition in the career of mind and especially significant for the view I am aiming to sketch. For thought, in thus becoming a conscious subject or object of thought, then began its destined course in reason: in ceasing to be merely an unconscious means of pursuit and becoming itself the quarry, it definitely entered upon the arduous way that leads to the goal of rigor. And so it is evident that the way in question is not a private way; it does not belong exclusively to mathematics; it is public property; it is the highway of conceptual research. For it is a mistake to imagine that mathematics, in virtue of its reputed exactitude, is an insulated science, dwelling apart in isolation from other forms and modes of conceptual activity. It would be such, were its rigor absolute; for between a perfection and any approximation thereto, however close, there always remains an infinitude of steps. But the rigor of mathematics is not absolute — absolute rigor is an ideal, to be, like other ideals, aspired unto, forever approached, but never quite attained, for such attainment would mean that every possibility of error or indetermination, however slight, had been eliminated from idea, from symbol, and from argumentation. We know, however, that such elimination can never be complete, unless indeed the human mind shall one day lose its insatiable faculty for doubting. What, then, *is* the distinction of mathematics on the score of exactitude? Its distinction lies, not in the attainment of rigor absolute, but partly in its exceptional devotion thereto and especially in the advancement it has made along the endless path that leads towards that perfection. But, as I have already said, it must not be thought that mathematics is the sole traveler upon the way. It is important to see clearly that it is far from being thus a solitary enterprise. First, however, let us adjust our imagery to a better correspondence with the facts. I have spoken of *the* path. We know, however, that the paths are many, as many as the varieties of conceptual subject-matter, all of them converging towards the same high goal. We see them originate here, there and yonder in the soil and haze of common thought; we see how indistinct they are at first — how ill-defined; we observe how they improve in that regard as the ideas involved grow clearer and clearer, more and more amenable to the use and governance of logic. At length, when thought, in its progress along any one of the many courses, has reached a high degree of refinement, precision and certitude, then and thereafter, but not before, we call it mathematical thought; it has undergone a long process of refining evolution and acquired at length the name of mathematics; it is not, however, the creature of its name; what is called mathematics has been long upon the way, owning at previous stages other designations — common sense, practical art perhaps, speculation, theology it may be, philosophy, natural science, or it may be for many a millennium no name at all. Is it, then, only a question of names? In a sense, yes: the ideal of thought is rigor; mathematics is the name that usage employs to designate, not attainment of the ideal, for it can not be attained, but its devoted pursuit and close approximation. But this is not the essence of the matter. The essence is that all thought, thought in all its stages, however rude, however refined, however named, owns the unity of being human: spiritual activities are one. Mathematics thus belongs to the great family of spiritual enterprises of man. These enterprises, all the members of the great family, however diverse in form, in modes of life, in methods of toil, in their progress along the way that leads towards logical rectitude, are alike children of one great passion. In genesis, in spirit and aspiration, in motive and aim, natural science, theology, philosophy, jurisprudence, religion and art are one with mathematics: they are all of them sprung from the human spirit's craving for invariant reality in a world of tragic change; they all of them aim at rescuing man from " the blind hurry of the universe from vanity to vanity " [7]: they seek cosmic stability — a world of abiding worth, where the broken promises of hope shall be healed and infinite aspiration shall cease to be mocked.

Such has been the universal and dominant aim and such are the cardinal forms that time has given its prosecution.

And now we must ask: What have been the fruits of the endless toil? What has the high emprize won? And what especially, have been the contributions of mathematics to the total gain? To recount the story of the spirit's quest for ageless forms of reality would be to tell afresh, from a new point of view, the history of human thought, so many and so diverse are the modes or aspects of being that men have found or fancied to be eternal. Edifying indeed would be the tale, but it is long, and the hour contracts. Even a meager delineation is hardly possible here. Yet we must not fail to glance at the endless array and to call, at least in part, the roll of major things. But where begin? Shall it be in theology? How memory responds to the magic word. " The past rises before us like a dream." [8] As the long succession of the theological centuries passes by, what a marvelous pageant do they present of human ideals, contrivings and dreams, both rational and superrational. Alpha and Omega, the beginning and ending, which is, which was and which is to come; I Am That I Am; Father of lights with which

---

[7] Bertrand Russell.   [8] Robert G. Ingersoll.

is no variableness, neither shadow of turning; the boni-
tas, unitas, infinitas, immutabilitas of Deity; the undying
principle of soul; the sublime hierarchy of immortal
angels, terrific and precious, discoursed of by sages, com-
memorated by art, feared and loved by millions of men
and women and children: these things may suffice to
remind us of the invariant forms of reality found or
invented by theology in her age-long toil and passion
to conquer the mutations of time by means of things
eternal.

But theology's record is only an immense chapter of
the vastly more inclusive annals of world-wide philosophic speculation running through the ages. If we turn
to philosophy understood in the larger sense, if we ask
what answers she has made in the long course of time to
the question of what is eternal, so diverse and manifold
are the voices heard across the centuries, from the East
and from the West, that the combined response must
needs seem to an unaccustomed ear like an infinite babel
of tongues: the Confucian Way of Heaven; the mystic
Tao, so much resembling fate, of Lao Tzu and Chuang
Tzu; Buddhism's inexorable spiritual law of cause and
effect and its everlasting extinction of individuality in
Nirvana — the final blowing out of consciousness and
character alike; Ahura Mazda, the holy One, of Zara-
thustra; Fate, especially in the Greek tragedies and Greek
religion — the chain of causes in nature, " the compulsion in the way things grow," a fine thread running
through the whole of existence and binding even the gods;
the cosmic matter, or το απειρον of Anaximander; the
cosmic order, the rhythm of events, the logos or reason
or nous, of Heracleitus; the finite, space-filling sphere, or
One, of the deep Parmenides; the four material and two
psychic, six eternal, elements of Empedocles; the infini-

tude of everlasting mind-moved simple substances of
Anaxagoras; the infinite multitude and endless variety
of invariant " seeds of things " of Leucippus, Democritus,
Epicurus and Lucretius, together with their doctrines of
absolute void and the conservations of mass and motion
and infinite room or space; Plato's eternal world of pure
ideas; the great Cosmic Year of a thousand thinkers,
rolling in vast endlessly repeated cycles on the beginningless, endless course of time from eternity to eternity; the
changeless thought-forms of Zeno, Gorgias and Aristotle;
Leibniz's indestructible, pre-established harmony; Spinoza's infinite unalterable substance; the Absolute of the
Hegelian school; and so on and on far beyond the limits
of practicable enumeration. This somewhat random partial list of things will serve to recall and to represent
the enormous motley crowd of answers that the ages of
philosophic speculation have made to the supreme inquiry
of the human spirit: what is there that survives the mutations of time, abiding unchanged despite the whirling
flux of life and the world?

And now, in the interest of further representing salient
features in a large perspective view, let me next ask what
contribution to the solution of the great problem has been
made by jurisprudence. Jurisprudence is no doubt at
once a branch of philosophy and a branch of science, but
it has an interest, a direction and a character of its own.
And for the sake of due emphasis it will be well worth
while to remind ourselves specifically of the half-forgotten
fact that, in its quest for justice and order among men,
jurisprudence long ago found an answer to our oft-stated
riddle of the world, an answer which, though but a partial
one, yet satisfied the greatest thinkers for many centuries,
and which, owing to the inborn supernalizing proclivity
of the human mind, still exercises sway over the thought

of the great majority of mankind. I allude to the conception of *jus naturale* or *lex naturæ*, the doctrine that in the order of Nature there somehow exists a perfect, invariant, universally and eternally valid system or prototype of law over and above the imperfect laws and changeful polities of men — a conception and doctrine long familiar in the juristic thought of antiquity, dominating, for example, the Antigone of Sophocles, penetrating the Republic and the Laws of Plato, proclaimed by Demosthenes in the Oration on the Crown, becoming, largely through the Republic and the Laws of Cicero, the crowning conception of the imperial jurisprudence of Rome, and still holding sway, as I have said, except in the case of our doubting Thomases of the law, who virtually deny our world the existence of any perfection whatever because they can not, so to speak, feel it with the hand, as if they did not know that to suppose an ideal to be *thus* realized would be a flat contradiction in terms.

If we turn for a moment to art and enquire what has been *her* relation to the poignant riddle, shall we not thus be going too far afield? The answer is certainly no. *In æternitatem pingo,* said Zeuxis, the Greek painter. " The purpose of art," says John La Farge, " is commemoration." In these two sayings, one of them ancient, the other modern, we have, I think, the evident clue. They do but tell us that art, like the other great enterprises of man, springs from our spirit's coveting of worth that abides. Like theology, like philosophy, like jurisprudence, like natural science, too, as I mean to point out further, and like mathematics, art is born of the universal passion for the dignity of things eternal. Her quest, like theirs, has been a search for invariants, for goods that are everlasting. And what has she found? The answer is simple. " The idea of beauty in each species of being,"

said Joshua Reynolds, " is perfect, invariable, divine."
We know that by a faculty of imaginative, mystical,
idealizing discernment there is revealed to us, amid the
fleeting beauties of Time, the immobile presence of
Eternal beauty, immutable archetype and source of the
grace and loveliness beheld in the shifting scenes of
the flowing world of sense. Such, I take it, is art's contribution to our human release from the tyranny of
change and the law of death.

And now what should be said of science? Not so brief
and far less simple would be the task of characterizing
or even enumerating the many things that in the great
drama of modern science have been assigned the rôle of
invariant forms of reality or eternal modes of being. It
would be necessary to mention first of all, as most imposing of all, our modern form of the ancient doctrine of
fate. I mean the reigning conception of our universe as
an infinite machine — a powerful conception that more
and more fascinates scientific minds even to the point of
obsession and according to which it should be possible,
were knowledge sufficiently advanced, to formulate, in a
system of differential equations, the whole of cosmic
history from eternity to eternity in minutest detail, not
even excluding a skeptic's doubt whether such formulation be theoretically possible nor excluding the conviction, which some minds have, that the doctrine, regarded
as an *ultimate* creed, is an abominable libel against the
character of a world where the felt freedom of the human
spirit is not an illusion. It would be necessary to mention — as next perhaps in order of impressiveness —
another doctrine that is, curiously enough, vividly remi-
niscent of old-time fate. I allude this time to the doctrine
of heredity, a tremendous conception, in accordance with
which — as Professor W. B. Smith has said in his recent

powerful address on " Push or Pull " ? — " the remotest past reaches out its skeletal fingers and grapples both present and future in its iron grip." And there is the conservation of energy and that of mass — both of them, again, doctrines prefigured in the thought of ancient Greece — and numerous other so-called natural laws, simple and complex, familiar and unfamiliar, all posing as permanent forms of reality — as natural invariants under the infinite system of cosmic transformations — and thus together constituting the enlarging contribution of natural science towards the slow vindication of a world that has seemed capricious, lawless and impermanent.

Such, then, is a conspectus, suggested rather than portrayed, of the results which the great allies of mathematics, operating through the ages, have achieved in their passionate endeavor to transcend the tragic vicissitudes and limitations of life in an " ever-growing and perishing " universe and to win at length the freedom, the dignity and the peace of a stable world where order and harmony reign and spiritual goods endure. If we are to arrive at a really just or worthy sense of the human significance of mathematics, it is in relation with those great results of her sister enterprises that the achievements of this science must be appraised. Immense indeed and high is the task of criticism as thus conceived. How diverse and manifold the doctrines to be evaluated, what depths to be plumbed, what heights to be scaled, how various the relationships and dignities to be assigned their rightful place in the hierarchy of values. In the presence of such a task what can we think or say in the remaining moments of the hour? If we have succeeded in setting the problem in its proper light and in indicating the sole eminence from which the matter may be rightly

viewed, we ought perhaps to be content with that as the
issue of the hour, for it is worth while to sketch a worthy
program of criticism even if time fails us to perform fully
the task thus set. And yet I can not refrain from inviting
you to imagine, before we close, a few at least of the
things that one who essayed the great critique would
submit to his auditors for meditation. And what do you
imagine the guiding lines and major themes of his discourse would be?

I fancy he would say: The question before us, ladies
and gentlemen, is not a question of weighing utilities nor
of counting applications nor of measuring material gains;
it is a question of human ideals together with the various
means of pursuing them and the differing degrees of their
approximation; we are occupied with a question of appreciation, with the problem of values. I am, he would say,
addressing you as representatives of man, and in so doing,
I am not regarding man as a mere practician, as a hewer
of wood and drawer of water, as an animal content to
serve the instincts for shelter and food and reproduction.
I am contemplating him as a spiritual being, as a thinker,
poet, dreamer, as a lover of knowledge and beauty and
wisdom and the joy of harmony and light, responding to
the lure of an ideal destiny, troubled by the mystery of
a baffling world, conscious subject of tragedy, yearning
for stable reality, for infinite freedom, for perpetuity and
a thousand perfections of life. As representatives of such
a being, you, he would say, and I, even if we be not ourselves producers of theology or philosophy or science or
jurisprudence or art or mathematics, are nevertheless
rightful inheritors of all this manifold wisdom of man.
The question is: What is the inheritance worth? We are
the heirs and we are to be the judges of the great responses that time has made to the spiritual needs of

humanity. What are the responses worth? What are their values, joint and several, absolute and relative? And what, especially, is the human worth of the response of mathematics? It is, he would say, not only our privilege, but, as educated individuals and especially as representatives of our race, it is our duty, to ponder the matter and reach, if we can, a right appraisement. For the proper study of mankind is man, and it is essential to remember that " *La vie de la science est la critique.*" [9] I have, he would say, tried to make it clear that mathematics is not an isolated science. I have tried to show that it is not an antagonist, or a rival, but is the comrade and ally of the other great forms of spiritual activity, all aiming at the same high end. I have reminded you of the principal answers made by these to the spiritual needs of man, and I do not, he would say, desire to underrate or belittle them. They are a precious inheritance. Many of them have not, indeed, stood the test of time; others will doubtless endure for aye; all of them, for a longer or shorter period, have softened the ways of life to millions of men and women. Neither do I desire, he would say, to exaggerate the contributions of mathematics to the spiritual weal of humanity. What I desire is a fair comparative estimate of its claims. " Truth is the beginning of every good thing, both to gods and men." I am asking you to compare, consider and judge for yourselves. The task is arduous and long.

There are, our critic would say, certain paramount considerations that every one in such an enterprise must weigh, and a few of them may, in the moments that remain, be passed in brief review. Consider, for example, our human craving for a world of stable reality. Where is it to be found? We know the answers of theology, of

[9] Cousin.

philosophy, of natural science and the rest. We know, too, the answer of literature and general thought:

> The cloud-capped towers, the gorgeous palaces,
> The solemn temples, the great globe itself,
> Yea, all which it inherit, shall dissolve,
> And, like the baseless fabric of this vision,
> Leave not a rack behind.

And now what, he would ask, is the answer of mathematics? The answer, he would have to say, is this: *Transcending the flux of the sensuous universe, there exists a stable world of pure thought, a divinely ordered world of ideas, accessible to man, free from the mad dance of time, infinite and eternal.*

Consider our human craving for freedom. Of freedom there are many kinds. Is it the freedom of limitless room, where our passion for outward expression, for externalization of thought, may attain its aim? It is to mathematics, our critic would say, that man is indebted for that priceless boon; for it is the cunning of this science that has at length contrived to release our long imprisoned thought from the old confines of our threefold world of sense and opened to its wing the interminable skies of hyperspace. But if it be a more fundamental freedom that is meant, if it be freedom of thought proper — freedom, that is, for the creative activity of intellect — then again it is to mathematics that our faculties must look for the definition and a right estimate of their prerogatives and power. For, regarding this matter, we may indeed acquire elsewhere a suspicion or an inkling of the truth, but mathematics, and nothing else, is qualified to give us *knowledge* of the fact that our intellectual freedom is absolute save for a single limitation — the law of non-contradiction, the law of logical compatibility, the law of intellectual harmony — sole restriction imposed by " the nature of things " or by logic or by the muses upon the creative activity of the human spirit.

Consider next, the critic might say, our human craving for a living sense of rapport and comradeship with a divine Being infinite and eternal. Except through the modern mathematical doctrine of infinity, there is, he would have to say, no rational way by which we may even approximate an understanding of the supernal attributes with which our faculty of idealization has clothed Deity — no way, except this, by which our human reason may gaze understandingly upon the downward-looking aspects of the overworld. But this is not all. I need not, he would say, remind you of the reverent saying attributed to Plato that " God is a geometrician." Who is so unfortunate as not to know something of the religious awe, the solace and the peace that come from cloistral contemplation of the purity and everlastingness of mathematical truth?

Mighty is the charm of those abstractions to a mind beset with images and haunted by himself.

" More frequently," says Wordsworth, speaking of geometry,

> More frequently from the same source I drew
> A pleasure quiet and profound, a sense
> Of permanent and universal sway,
> And paramount belief; there, recognized
> A type, for finite natures, of the one
> Supreme Existence, the surpassing life
> Which to the boundaries of space and time,
> Of melancholy space and doleful time,
> Superior and incapable of change,
> Not touched by welterings of passion — is,
> And hath the name of God. Transcendent peace
> And silence did wait upon those thoughts
> That were a frequent comfort to my youth.

And so our spokesman, did time allow, might continue, inviting his auditors to consider the relations of mathematics to yet other great ideals of humanity — our human craving for rectitude of thought, for ideal justice, for dominion over the energies and ways of the material universe, for imperishable beauty, for the dignity and peace of intellectual harmony. We know that in all such cases the issue of the great critique would be the same, and it is needless to pursue the matter further. The light is clear enough. Mathematics is, in many ways, the most precious response that the human spirit has made to the call of the infinite and eternal. It is man's best revelation of the " Deep Base of the World." [10]

[10] Sir Gilbert Murray.

THE HUMANIZATION OF THE TEACHING OF MATHEMATICS [1]

WHEN the distinguished chairman of your mathematical conference did me the honor to request me to speak to you, he was generous enough, whether wisely or unwisely, to leave the choice of a subject to my discretion, merely stipulating that, whatever the title might be, the address itself should bear upon the professional function of those men and women who are engaged in teaching mathematics in secondary schools. Inexpertness, it has been said, is the curse of the world; and one may, not unnaturally, feel some hesitance in undertaking a task that might seem to resemble the rôle of a physician when, as sometimes happens, he is called upon to treat a patient whose health and medical competence surpass his own. I trust I am not wanting in that natural feeling. In the present instance two considerations have enabled me to overcome it. One of them is that, having had some experience in teaching mathematics in secondary schools, I might, it seemed to me, regard that experience, though it was gained more than a score of years ago, as giving something like a title to be heard in your counsels. The other consideration is that, in regard to the teaching of mathematics, whether in secondary schools or in colleges, I have acquired a certain conviction, a pretty firm conviction, which, were it properly presented, you would

[1] Address given at the meeting of the Michigan School Masters' Club, at Ann Arbor, March 28, 1912. Printed in *Science*, April 26, 1912; in *The Educational Review*, September, 1912; and in the *Michigan School Masters' Magazine*.

doubtless be generous enough and perhaps ingenious enough to regard as having some sort of likeness to a message.

My conviction is, that hope of improvement in mathematics teaching, whether in secondary schools or in colleges, lies mainly in the possibility of humanizing it. It is worth while to remember that our pupils are human beings. What it means to be a human being we all of us presumably know pretty well; indeed we know it so well that we are unable to tell it to one another adequately; and, just because we do so well know what it means to be a human being, we are prone to forget it as we forget, except when the wind is blowing, that we are constantly immersed in the earth's atmosphere. To humanize the teaching of mathematics means so to present the subject, so to interpret its ideas and doctrines, that they shall appeal, not merely to the computatory faculty or to the logical faculty but to all the great powers and interests of the human mind. That mathematical ideas and doctrines, whether they be more elementary or more advanced, admit of such a manifold, liberal and stimulating interpretation, and that therefore the teaching of mathematics, whether in secondary schools or in colleges, may become, in the largest and best sense, human, I have no doubt. That mathematical ideas and doctrines do but seldom receive such interpretation and that accordingly the teaching of mathematics is but seldom, in the largest and best sense, human, I believe to be equally certain. That the indicated humanization of mathematical teaching, the bringing of the matter and the spirit of mathematics to bear, not merely upon certain fragmentary faculties of the mind, but upon the whole mind, that this is a great desideratum is, I assume, beyond dispute.

How can such humanization be brought about? The answer, I believe, is not far to seek. I do not mean that the answer is easy to discover or easy to communicate. I mean that the game is near at hand and that it is not difficult to locate it, though it may not be easy to capture it. The difficulty inheres, I believe, in our conception of mathematics itself; not so much in our conception of what mathematics, in a definitional sense, is, for that sense of what mathematics is has become pretty clear in our day, but in our sense or want of sense of what mathematics, whatever it may be, humanly signifies. In order to humanize mathematical teaching it is necessary, and I believe it is sufficient, to come under the control of a right conception of the human significance of mathematics. It is sufficient, I mean to say, and it is necessary, greatly to enlarge, to enrich and to vitalize our sense of what mathematics, regarded as a human enterprise, signifies.

What does mathematics, regarded as an enterprise of the human spirit, signify? What is a just and worthy sense of the human significance of mathematics?

To the extent in which any of us really succeeds in answering that question worthily, his teaching will have the human quality, in so far as his teaching is, in point of external circumstance, free to be what it would. I believe it is important to put the question, and it is with the putting of it rather than with the proposing of an answer to it that I am here at the outset mainly concerned. For any one who is really to acquire possession of an answer that is worthy must win the answer for himself. I need not say to you that such an acquisition as a worthy answer to this kind of question does not belong to the category of things that may be lent or borrowed, sold or bought, donated or acquired by gift.

No doubt the answers we may severally win will differ as our temperaments differ. Yet the matter is not solely a matter of temperament. It is much more a matter first of knowledge and then of the evaluation of the knowledge and of its subject. To the winning of a worthy' sense of the human significance of mathematics two things are indispensable, knowledge and reflection: knowledge of mathematics and reflection upon it. To the winning of such a sense it is essential to have the kind of knowledge that none but serious students of mathematics can gain. Equally essential is another thing and this thing students of mathematics in our day do not, or do but seldom, gain. I mean the kind of insight and the liberality of view that are to be acquired only by prolonged contemplation of the nature of mathematics and by prolonged reflection upon its relations of contrast and similitude to the other great forms of spiritual activity.

The question, though it is a question about mathematics, is not a mathematical question; it is a philosophical question. And just because it is a philosophical question, mathematicians, despite the fact that one of the indispensable qualifications for considering it is possessed by them alone, have in general ignored it. They have, in general, ignored it, and their ignoring of it may help to explain the curious paradox that whilst the world, whose mathematical knowledge varies from little to less, has always as if instinctively held mathematical science in high esteem, it has at the same time usually regarded mathematicians as eccentric and abnormal, as constituting a class apart, as being something more or something less than human. It may explain, too, I venture to believe it does partly explain, both why it is that in the universities the number of students attracted to advanced lectures in mathematics compared with the numbers drawn to advanced courses in some other great subjects not inherently more attractive, is so small; and why it is that, among the multitudes who pursue mathematics in the secondary schools, only a few find in the subject anything like delight. For I do not accept the traditional and still current explanation, that the phenomenon is due to a well-nigh universal lack of mathematical faculty. I maintain, on the contrary, that a vast majority of mankind possess mathematical faculty in a very considerable degree. That the average pupil's interest in mathematics is but slight, is a matter of common knowledge. His lack of interest is, in my opinion, due, not to a lack of the appropriate faculty in him, but to the circumstance that he is a human being, whilst mathematics, though it teems with human interest, is not presented to him in its human guise.

If you ask the world — represented, let us say, by the man in the street or in the market place or the field — to tell you its estimate of the human significance of mathematics, the answer of the world will be, that mathematics has given mankind a metrical and computatory art essential to the effective conduct of daily life, that mathematics admits of countless applications in engineering and the natural sciences, and finally that mathematics is a most excellent instrumentality for giving mental discipline. Such will be the answer of the world. The answer is intelligible, it is important, and it is good so far as it goes; but it is far from going far enough and it is not intelligent. That it is far from going far enough will become evident as we proceed. That the answer is not intelligent is evident at once, for the first part of it seems to imply that the rudimentary mathematics of the carpenter and the counting-house is scientific, which it is not; the second part of the answer is but an echo by

the many of the voice of the few; and, as to the final
part, the world's conception of intellectual discipline is
neither profound nor well informed but is itself in sorry
need of discipline.

If, turning from the world to a normal mathematician,
you ask him to explain to you the human significance of
mathematics, he will repeat to you the answer of the
world, of course with far more appreciation than the
world has of what the answer means, and he will supplement the world's response by an important addition. He
will add, that is, that mathematics is the exact science,
the science of exact thought or of rigorous thinking. By
this he will not mean what the world would mean if the
world employed, as sometimes it does employ, the same
form of words. He will mean something very different.
Especially if he be, as I suppose him to be, a normal
mathematician of the modern critical type, he will mean
that mathematics is, in the oft-cited language of Benjamin
Peirce, " the science that draws necessary conclusions; "
he will mean that, in the felicitous words of William
Benjamin Smith, " mathematics is the universal art apodictic; " he will mean that mathematics is, in the nicely
technical phrase of Pieri, " a hypothetico-deductive system." If you ask him whether mathematics is the science
of rigorous thinking about *all* the things that engage the
thought of mankind or only about a few of them, such
as numbers, figures, certain operations, and the like, the
answer he will give you depends. If he be a normal
mathematician of the elder school, he will say that mathematics is the science of rigorous thinking about only a
relatively few things and that these are such as you have
exemplified. And if now, with a little Socratic persistence, you press him to indicate the human significance
of a science of rigorous thinking about only a few of the

countless things that engage human thought, his answer
will give you but little beyond a repetition of the above-
mentioned answer of the world.   But if he be a normal
mathematician of the modern critical type, he will say
that mathematics is the science of rigorous thinking
about all the things that engage human thought, about *all*
of them, he will mean, in the sense that thinking, as it
approaches perfection, tends to assume certain definite
forms, that these forms are the same whatever the subject
matter of the thinking may be, and that mathematics is
the science of these forms *as forms*.   If you respond, as
you well may respond, that, in accordance with this ontological conception of mathematics, this science, instead
of thinking about *all*, thinks about *none*, of the concrete
things of interest to human thought, and that accordingly
Mr. Bertrand Russell was right in saying that " mathematics is the science in which one never knows what one
is talking about nor whether what one says is true " —
if you respond that, from the point of view above
assumed, that delicious *mot* of Mr. Russell's must be
solemnly held as true, and then if, in accordance with
your original purpose, you once more press for an estimation of the human significance of such a science, I
fear that the reply, if your interlocutor is a mathematician of the normal type, will contain little that is new
beyond the assertion that the science in question is very
interesting, where, by interesting, he means, of course,
interesting to mathematicians.   It is true that Professor
Klein has said: " Apart from the fact that pure mathematics can not be supplanted by anything else as a means
for developing the purely logical faculties of the mind,
there must be considered here as elsewhere the necessity
of the presence of a few individuals in each country
developed in a far higher degree than the rest, for the

purpose of keeping up and gradually raising the *general* standard. Even a slight raising of the general level can be accomplished only when some few minds have progressed far ahead of the average." Here indeed we have, in these words of Professor Klein, a hint, if only a hint, of something better. But Professor Klein is not a mathematician of the normal type, he is hypernormal. If, in order to indicate the human significance of mathematics regarded as the science of the forms of thought as forms, your normal mathematician were to say that these forms constitute, of themselves, an infinite and everlasting world whose beauty, though it is austere and cold, is pure, and in which is the secret and citadel of whatever order and harmony our concrete universe contains, it would yet be your right and your duty to ask, as the brilliant author of " East London Visions " once asked me, namely, what is the human significance of " this majestic intellectual cosmos of yours, towering up like a million-lustered iceberg into the arctic night," seeing that, among mankind, none is permitted to behold its more resplendent wonders save the mathematician himself?   But the normal mathematician will not say what I have just now supposed him to say; he will not say it, because he is, by hypothesis, a normal mathematician, and because, being a normal mathematician, he is exclusively engaged in exploring the iceberg.   A farmer was once asked why he raised so many hogs.   " In order," he said, " to buy more land."   Asked why he desired more land, his answer was, " in order to raise more corn."   Being asked to say why he would raise more corn, he replied that he wished to raise more hogs.   If you ask the normal mathematician why he explores the iceberg so much, his answer will be, in effect at least, " in order to explore it more."   In this exquisite circularity of motive, the farmer and the normal mathematician are well within their rights. They are within their rights just as a musician would be within his rights if he chanced to be so exclusively interested in the work of composition as never to be concerned with having his creations rendered before the public and never to attempt a philosophic estimate of the human worth of music. The distinction involved is not the distinction between human and inhuman, between social and anti-social; it is the distinction between what is human *or* inhuman, social or anti-social, and what is neither the one nor the other. No one, I believe, may contest the normal mathematician's right as a mathematical student or investigator to be quite indifferent as to the social value or the human worth of his activity. Such activity is to be prized just as we prize any other natural agency or force that, however undesignedly, yet contributes, sooner or later, directly or indirectly, to the weal of mankind. The fact is that, among motives in research, scientific curiosity, which is neither moral nor immoral, is far more common and far more potent than charity or philanthropy or benevolence. But when the mathematician passes from the rôle of student or investigator to the rôle of teacher, that right of indifference ceases, for he has passed to an office whose functions are social and whose obligations are human. It is not his privilege to chill and depress with the encasing fogs of the iceberg. It is his privilege and his duty, in so far as he may, to disclose its " million-lustered " splendors in all their power to quicken and illuminate, to charm and edify, the whole mind.

The conception of mathematics as the science of the forms of thought as forms, the conception of it as the refinement, prolongation and elaboration of pure logic, is, as you are doubtless aware, one of the great out-

comes, perhaps I should say it is the culminating philosophical outcome, of a century's effort to ascertain what
mathematics, in its intimate structure, is. This conception of what mathematics is comes to its fullest expression and best defense, as you doubtless know, in such
works as Schroeder's " Algebra der Logik," Whitehead's
" Universal Algebra," Russell's " Principles of Mathematics," Peano's " Formulario Matematico," and especially in Whitehead and Russell's monumental " Principia
Mathematica." I cite this literature because it tells us
what, in a definitional sense, the science in which the
normal mathematician is exclusively engaged, is. If we
wish to be told what that science humanly signifies, we
must look elsewhere; we must look to a mathematician
like Plato, for example, or to a philosopher like Poincaré,
but especially must we look to our own faculty for discerning those fine connective things — community of aim,
interformal analogies, structural similitudes — that bind
all the great forms of human activity and aspiration —
natural science, theology, philosophy, jurisprudence, religion, art and mathematics — into one grand enterprise
of the human spirit.

In the autumn of 1906 there was published in *Poet
Lore* a short poem which, though it says nothing explicitly
of mathematics, yet admits of an interpretation throwing
much light upon the human significance of the science
and indicating well, I think, the normal mathematician's
place in the world of spiritual interests. The author of
the poem is my excellent friend and teacher, Professor
William Benjamin Smith, mathematician, philosopher,
poet and theologian. I have not asked his permission to
interpret the poem as I shall invite you to interpret it.
What its original motive was I am not informed — it may
have been the exceeding beauty of the ideas expressed in

it or the harmonious mingling of their light with the
melody of their song. The title of the poem is " The
Merman and the Seraph." As you listen to the reading
of it, I shall ask you to regard the Merman as representing the normal mathematician and the Seraph as representing, let us say, the life of the emotions in their
higher reaches and their finer susceptibilities.

> Deep the sunless seas amid,
> Far from Man, from Angel hid,
> Where the soundless tides are rolled
> Over Ocean's treasure-hold,
> With dragon eye and heart of stone,
> The ancient Merman mused alone.
>
> And aye his arrowed Thought he wings
> Straight at the inmost core of things —
> As mirrored in his Magic glass
> The lightning-footed Ages pass, —
> And knows nor joy nor Earth's distress,
> But broods on Everlastingness.
> " Thoughts that love not, thoughts that hate not,
> Thoughts that Age and Change await not,
>   All unfeeling,
>   All revealing,
>   Scorning height's and depth's concealing,
> These be mine — and these alone! " —
> Saith the Merman's heart of stone.
>
> Flashed a radiance far and nigh
> As from the vertex of the sky, —
> Lo! a Maiden beauty-bright
> And mantled with mysterious might
> Of every power, below, above,
> That weaves resistless spell of Love.
>
> Through the weltering waters cold
> Shot the sheen of silken gold;
> Quick the frozen Heart below
> Kindled in the amber glow;
> Trembling Heavenward Nekkan yearned
> Rose to where the Glory burned.

" Deeper, bluer than the skies are,
Dreaming meres of morn thine eyes are
    All that brightens
    Smile or heightens
    Charm is thine, all life enlightens,
Thou art all the soul's desire." —
Sang the Merman's Heart of Fire.

" Woe thee, Nekkan! Ne'er was given
Thee to walk the ways of Heaven;
    Vain the vision,
    Fate's derision,
    Thee that raps to realms elysian,
Fathomless profounds are thine " —
Quired the answering voice divine.

Came an echo from the West,
Pierced the deep celestial breast;
Summoned, far the Seraph fled,
Trailing splendors overhead;
Broad beneath her flying feet
Laughed the silvered ocean-street.

On the Merman's mortal sight
Instant fell the pall of Night;
Sunk to the sea's profoundest floor
He dreams the vanished Vision o'er,
Hears anew the starry chime,
Ponders aye Eternal Time.
" Thoughts that hope not, thoughts that fear not,
Thoughts that Man and Demon veer not
    Times unending
    Comprehending,
    Space and worlds of worlds transcending,
These are mine — but these alone! " —
Sighs the Merman's heart of stone.

I have said that the poem, if it receive the interpretation that I have invited you to give it, throws much light on the human significance of mathematics and indicates well the place of the normal mathematician in the world of spiritual interests. No doubt the place of the

Merman and the place of the Angel are not the same: no doubt the world of whatsoever in thought is passionless, infinite and everlasting, and the world of whatsoever in feeling is high and beauteous and good are distinct worlds, and they are sundered wide in the poem. But, though in the poem they are held widely apart, in the poet they are united. For the song is not the Merman's song nor are its words the words of the Seraph. It is the voice of the poet — a voice of man. The Merman's world and the world of the Seraph are not the same, they are very distinct; in conception they are sundered; they may be sundered in life, but in life it need not be so. The Merman indeed is confined to the one world and the Seraph to the other, but man, a man unless he be a Merman, may inhabit them both. For the angel's denial, the derision of fate, is not spoken of man, it is spoken of the Merman; and the Merman's sigh is not his own, it is a human sigh — so lonely seems the Merman in the depths of his abode.

No, the world of interests of the human spirit is not the Merman's world alone nor the Seraph's alone. It is not so simple. It is rather a cluster of worlds, of worlds that differ among themselves as differ the lights by which they are characterized. As differ the lights. The human spirit is susceptible of a variety of lights and it lives at once in a corresponding variety of worlds. There is perception's light, commonly identified with solar radiance or with the radiance of sound, for music, too, is, to the spirit, a kind of illumination: perceptional light, in which we behold the colors, forms and harmonies of external nature: a beautiful revelation—a world in which any one might be willing to spend the remainder of his days if he were but permitted to live so long. And there is imagination's light, disclosing a new world filled with wondrous things, things that may or may not resemble the things revealed in perception's light but are never identical with them: light that is not superficial nor constrained to paths that are straight but reveals the interiors of what it illuminates and phases that look away. Again, there is the light of thought, of reason, of logic, the light of analysis, far dimmer than perception's light, dimmer, too, than that of imagination, but far more penetrating and far more ubiquitous than either of them, disclosing things that curiously match the things that they disclose and countless things besides, namely, the world of ideas and the relations that bind them: a cosmic world, in the center whereof is the home of the Merman. There remains to be named a fourth kind of light. I mean the light of emotion, the radiance and glory of things that, save by gleams and intimations, are not revealed in perception or in imagination or in thought: the light of the Seraph's world, the world of the good, the true and the beautiful, of the spirit of art, of aspiration and of religion.

Such, in brief, is the cluster of worlds wherein dwell the spiritual interests of the human beings to whom it is our mission to teach mathematics. My thesis is that it is our privilege to show, in the way of our teaching it, that its human significance is not confined to one of the worlds but, like a subtle and ubiquitous ether, penetrates them all. Objectively viewed, conceptually taken, these worlds, unlike the spheres of the geometrician, do not intersect — a thing in one of them is not in another; but the things in one of them and the things in another may own a fine resemblance serving for mutual recall and illustration, effecting transfer of attention — transformation as the mathematicians call it — from world to world; for whilst these worlds of interest, objectively viewed, have naught in common, yet subjectively they are united,

united as differing mansions of the house of the human
spirit. A relation, for example, between three independent variables exists only in the grey light of thought, only
in the world of the Merman; the habitation of the geometric locus of the relation is the world of imagination;
if a model of the locus be made or a drawing of it, this
will be a thing in the world of perception; finally, the
wondrous correlation of the three things, or the spiritual
qualities of them — the sensuous beauty of the model or
the drawing, the unfailing validity of the given relation
holding as it does throughout " the cycle of the eternal
year," the immobile presence of the locus or image poised
there in eternal calm like a figure of justice — these may
serve, in contemplating them, to evoke the radiance of
the Seraph's world: and thus the circuit and interplay,
ranging through the world of imagination and the world
of thought from what is sensuous to what is supernal, is
complete. It would not have seemed to Plato, as it may
seem to us, a far cry from the prayer of a poet to the
theorem of Pythagoras, for example, or to that of Archimedes respecting a sphere and its circumscribing cylin-
der. Yet I venture to say, that calm reflection upon the
existence and nature of such a theorem — cloistral contemplation, I mean, of the fact that it is really true, of its
serene beauty, of its silent omnipresence throughout the
infinite universe of space, of the absolute exactitude and
invariance of its truth from everlasting to everlasting —
such reflection will not fail to yield a sense of reverence
and awe akin to the feeling that, for example, pervades
this choral prayer by Sophocles:
" Oh! that my lot may lead me in the path of holy
innocence of word and deed, the path which august laws
ordain, laws that in the highest empyrean had their birth,
of which Heaven is the father alone, nor did the race of

mortal men beget them, nor shall oblivion put them to sleep. The god is mighty in them and he groweth not old."

But why should we think it strange that interests, though they seem to cluster about opposite poles, are yet united by a common mood? Of the great world of human interests, mathematics is indeed but a part; but it is a central part, and, in a profound and precious sense, it is " the eternal type of the wondrous whole." For poetry and painting, sculpture and music — art in all its forms — philosophy, theology, religion and science, too, however passional their life and however tinged or deeply stained by local or temporal circumstance, yet have this in common: they all of them aim at values which transcend the accidents and limitations of every time and place; and so it is that the passionlessness of the Merman's thought, the infiniteness of the kind of being he contemplates and the everlastingness of his achievements enter as essential qualities into the ideals that make the glory of the Seraph's world. I do not forget, in saying this, that, of all theory, mathematical theory is the most abstract. I do not forget that mathematics therefore lends especial sharpness to the contrast in the Mephistophelian warning:

> Grey, my dear friend, is all theory,
> Green the golden tree of life.

Yet I know that one who loves not the grey of a naked woodland has much to learn of the esthetic resources of our northern clime. A mathematical doctrine, taken in its purity, is indeed grey. Yet such a doctrine, a world-filling theory woven of grey relationships finer than gossamer but stronger than cables of steel, leaves upon an intersecting plane a tracery surpassing in fineness and beauty the exquisite artistry of frost-work upon a windowpane. Architecture, it has been said, is frozen music. Be it so. Geometry is frozen architecture.

No, the belief that mathematics, because it is abstract, because it is static and cold and grey, is detached from life, is a mistaken belief. Mathematics, even in its purest and most abstract estate, is not detached from life. It is just the ideal handling of the problems of life, as sculpture may idealize a human figure or as poetry or painting may idealize a figure or a scene. Mathematics is precisely the ideal handling of the problems of life, and the central ideas of the science, the great concepts about which its stately doctrines have been built up, are precisely the chief ideas with which life must always deal and which, as it tumbles and rolls about them through time and space, give it its interests and problems, and its order and rationality. That such is the case a few indications will suffice to show. The mathematical concepts of constant and variable are represented familiarly in life by the notions of fixedness and change. The concept of equation or that of an equational system, imposing restriction upon variability, is matched in life by the concept of natural and spiritual law, giving order to what were else chaotic change and providing partial freedom in lieu of none at all. What is known in mathematics under the name of limit is everywhere present in life in the guise of some ideal, some excellence high-dwelling among the rocks, an " ever flying perfect " as Emerson calls it, unto which we may approximate nearer and nearer, but which we can never quite attain, save in aspiration. The supreme concept of functionality finds its correlate in life in the all-pervasive sense of interdependence and mutual determination among the elements of the world. What is known in mathematics as transformation — that is, lawful transfer of attention, serving to match in orderly fashion the things of one system with those of another — is conceived in life as a process of transmutation by which, in the flux of the world, the content of the present has come out of the past and in its turn, in ceasing to be, gives birth to its successor, as the boy is father to the man and as things, in general, become what they are not. The mathematical concept of invariance and that of infinitude, especially the imposing doctrines that explain their meanings and bear their names — what are they but mathematicizations of that which has ever been the chief of life's hopes and dreams, of that which has ever been the object of its deepest passion and of its dominant enterprise, I mean the finding of worth that abides, the finding of permanence in the midst of change, and the discovery of the presence, in what has seemed to be a finite world, of being that is infinite?   It is needless further to multiply examples of a correlation that is so abounding and complete as indeed to suggest a doubt which is the juster, to view mathematics as the abstract idealization of life, or to regard life as the concrete realization of mathematics.

Finally, I wish to emphasize the fact that the great concepts out of which the so-called higher mathematical branches have grown — the concepts of variable and constant, of function, class and relation, of transformation, invariance, and group, of finite and infinite, of discreteness, limit, and continuity — I wish, in closing, to emphasize the fact that these great ideas of the higher mathematics, besides penetrating life, as we have seen, in all its complexity and all its dimensions, are omnipresent, from the very beginning, in the *elements* of mathematics as well.   The notion of group, for example, finds easy and beautiful illustration, not only among the

simpler geometric notions and configurations, but even in
the ensemble of the very integers with which we count.
The like is true of the distinction of finite and infinite,
and of the ideas of transformation, of invariant, and
nearly all the rest.[2] Why should the presentation of them
have to await the uncertain advent of graduate years of
study? For life already abounds, and the great ideas
that give it its interests, order and rationality, that is to
say, the focal concepts of the higher mathematics, are
everywhere present in the elements of the science as
glistening bassets of gold. It is our privilege, in teaching
the elements, to avail ourselves of the higher conceptions
that are present in them; it is our privilege to have and
to give a lively sense of their presence, their human significance, their beauty and their light. I do not advocate
the formal presentation, in secondary schools, of the
higher conceptions, in the way of printed texts, for the
printed text is apt to be arid and the letter killeth. What
I wish to recommend is the presentation of them, as
opportunity may serve, in Greek fashion, by means of
dialectic, face to face, voice answering to voice, animated
with the varying moods and motions and accents of life
— laughter, if you will, and the lightning of wit to cheer
and speed the slower currents of sober thought. Of dialectic excellence, Plato at his best, as in the " Phædo "
or the " Republic," gives us the ideal model and eternal
type. But Plato's ways are frequently circuitous, wearisome and long. They are ill suited to the manners of a
direct and undeliberate age; and we must find, each for
himself, a shorter course. Somebody imbued with the
spirit of the matter, possessed of ample knowledge and

[2] The layman will find the great pillar concepts of mathematics
simply explained in Keyser's *Mathematical Philosophy*, E. P. Dutton &
Company.

having, besides, the requisite skill and verve ought to write a book showing, in so far as the printed page can be made to show, how naturally and swiftly and with what a delightful sense of emancipation and power thought may pass by dialectic paths from the traditional elements of mathematics both to its larger concepts and to a vision of their bearings on the higher interests of life. I need not say that such a handling of ideas implies much more than a verbal knowledge of their definitions. It implies familiarity with the doctrines that unfold the meanings of the ideas defined. It is evident that, in respect of this matter, the scripture must read: Knowing the doctrine is essential to living the life.

THE WALLS OF THE WORLD: OR CONCERNING THE FIGURE AND THE DIMENSIONS OF THE UNIVERSE OF SPACE [1]

THERE is something a little incongruous in attempting to consider the subject of this address in a theater or lecture hall whose roof and walls shut out from view the wide expanses of the world and the azure deeps. For how can we, amid the familiar finite scenes of a closed and blinded room, command a fitting mood for contemplating the infinite scenes without and beyond? A subject that has sheer vastness for its central or major theme demands for its appropriate contemplation the still expanse of some vast and open solitude, such as the peak of a lone and lofty mountain would afford, where the gaze meets no wall save the far horizon and no roof but the starry sky. Perhaps you will be good enough for the time to transport yourselves, in imagination, into the stillness of such a solitude, so that in the musing spirit of the place the questions to be propounded for consideration here may arise naturally and give us a due sense of their significance and impressiveness. What are the dimensions and what is the figure of our universe of space? How big is it and what is its shape? What is the figure of it and what is its size?

[1] An address delivered under the auspices of the local chapters of the Society of Sigma Xi at the state universities of Minnesota, Nebraska and Iowa, April 24, 28 and 30, 1913, respectively, and at a joint meeting of the chapters of Sigma Xi and Phi Beta Kappa of Columbia University, May 8, 1913. Printed in *Science,* June 13, 1913.

I do not mind owning that these questions have haunted me a good deal from the days of my youth. It happened in those days, though I was not aware of it nor became aware of it till after many years, that there were then coming into mathematics, just entering the fringe, so to speak, or the vestibule of the science, certain striking ideas which, as I venture to hope we may see, were destined, if not indeed to enable us to answer the questions with certainty, at all events to clarify them, to enrich their meaning and to make it possible to discuss them profitably. It has not been my fortune to meet many persons who had seriously propounded the questions to themselves or who seemed to be immediately interested in them when propounded by others — not many, even among astronomers, whose minds, it may be assumed, are especially " accustomed to contemplation of the vast." And so I have been forced to the somewhat embarrassing conclusion that my own long interest in the questions has been due to the fact of my being of a specially practical turn of mind. Quite seriously I venture to say that we are here engaged in a practical enterprise. For even if the questions were in the nature of the case unanswerable, which we do not admit, who does not know how great the boons that have come to men through pursuit of the unattainable? And who does not know that, as Mr. Chesterton has said, if you wish really to know a man, the most practical question to ask is, not about his occupation or his club membership or his party or church affiliations, but what are his views of the all-embracing world? What does he think of the universe? Do but fancy for a moment that in somewise men should come to *know* the exact shape or figure and especially the exact size or dimensions of the all-immersing space of our universe. It requires but little imagina-

tion, not much reflection, no extensive knowledge of cos-
mogonic history and speculation, no very profound insight into the ways of truth to men, it needs, I say, but
little philosophic sense to see that such knowledge would
in a thousand ways, direct and indirect, react powerfully
upon our whole intelligence, upon all our attitudes, sentiments and views, transforming our theology, our ethics,
our art, our religion, our philosophy, our literature, our
science, and therewith affecting profoundly the whole
sense and manner, the tone, color and meaning, of all our
institutions and the affairs of daily life. Nothing is quite
so practical, in the sense of being effectual and influential, as the views men hold, consciously or unconsciously,
regarding the great locus of their lives and their cosmic
home.

In order to discuss the questions before us intelligibly
and profitably it is not necessary by way of clearing the
ground to enter far into metaphysical speculation or into
psychological analysis with a view to ascertaining what
it is that we mean or ought to mean by space. We are
not obliged to dispute, much less decide, whether space
is subjective or objective or both or indeed something
that, as Plato in the " Timæus " acutely contends, is
neither the one nor the other. We may or may not agree
with the contention of Kant that space is, not an object,
but the form, of outer sense; we may or may not agree
with the radically different contention of Poincaré that
(geometric as distinguished from sensible) space is nothing but what is known in mathematics as a group, of
which the concept " is imposed on us, not as form of our
sense, but as form of our understanding." It is, I say,
not necessary for us, in the interest of soundness and
intelligibility, to try to compose such differences or to
attempt a settlement of these profound and important

questions. As to the distinction between sensible space
and geometric space, it would indeed be indispensable to
draw it sharply and to keep it always in mind, if we were
undertaking to ascertain what the subject (or the object)
of geometry is, or, what is tantamount, if we were seeking to get clearly aware of what it is that geometry is
about. But in discussing the subject before us it is unnecessary to be always guarding that distinction; for,
whilst it is the space of geometry, and not sensible space,
that we shall be talking about, yet it would be a hin-
drance rather than a help if we did not allow, as we
habitually do allow, the two varieties of space — the
imagery of the one, the conceptual character of the other
— to mingle freely in our thinking. There will be finesse
enough for the keenest arrows of our thought without
our going out of the way to find it. A procedure less
sophisticated will suffice. It will be sufficient to regard
space as being what, to the layman and to the student
of natural science, it has always seemed to be: a vast
region or room round about us, an immense exteriority,
locus of all suspended and floating objects of outer sense,
the whence, where and whither of motion, theater, in a
word, of the ageless drama of the physical universe. In
naturally so construing the term we do not commit ourselves to the philosophy, so-called, of common sense; we
thus merely save our discourse from the encumbrance of
needless refinements; for it is obvious that, if space be
not indeed what we have said it seems to be, the seeming
is yet a fact, and our questions would remain without
essential change: what, then, we should ask, are the
dimensions and what is the figure of that seeming?

Though all the things contained within that triply
extended spread or expanse which we call space are subject to the law of ceaseless change, the expanse itself, the

container of all, appears to suffer no variation whatever, but to be, unlike time, a genuine constant, the same yesterday, today and forever, sole absolute invariant under the infinite host of transformations that constitute the cosmic flux. Whether it be so in fact, of course we do not know. We only know that no good reason has ever been advanced for holding the contrary as an hypothesis.

And yet there is a sense, which we ought I think to notice, an interesting sense, in which space seems to be, not a constant, but, like time, a variable. There is a sense, deeper and juster perhaps than at first we suspect, in which the space of our universe has in the course of time alternately shrunken and grown. During the last century, for example, it has, so it seems, greatly grown, in response, it may be, to an increasing need of the human mind. By grown I do not mean grown in the usual sense, I do not mean the biological sense, I do not mean the sense that was present to the mind of that great man Leonardo da Vinci, when he wrote in effect as follows: if you wish to know that the earth has been growing, you have only to observe " how, among the high mountains, the walls of ancient and ruined cities are being covered over and concealed by the earth's increase "; and, if you would learn how *fast* the earth is growing, you have only to set a vase, filled with pure earth, upon a roof; to note how green herbs will immediately begin to shoot up; to note that these, when mature, will cast their seeds; to allow the process to continue through repetition; then, after the lapse of a decade, to measure the soil's increase; and, finally, to multiply, in order to have thus determined " how much the earth has grown in the course of a thousand years." In this matter, Leonardo was doubtless wrong. At all events current scientific views are against him. The earth, we know, has grown, but the growth has been by accretion, by addition from without, and not, in biologic sense, by expansion from within (unless, indeed, we adopt the beautiful hypothesis of the poet and physicist, Theodor Fechner, for which so hard-headed a scientific man as Bernhardt Riemann had so much respect, the hypothesis, namely, that the plants, the earth and the stars have souls). The myriad-minded Florentine was, we of today think, in error, his error being one of those brilliant mistakes that but few men have been qualified to make. But in saying that space has grown we do not mean that it has grown in the biologic sense of Leonardo nor yet in the sense of addition from without. We mean that it has grown as a thing in mind may grow, as a thing in thought may grow; we mean that it has grown in men's conception of it. That space has, in this sense, been enlarged pro-digiously in the course of recent time is evident to all. It has been often said that " the first grand discovery of modern times is the immense extension of the universe *in space*." [2] It would be juster to say that the first grand achievement of modern science has been the immense extension of space itself, the prodigious enlargement of it, in the imagination and especially in the thought of men. If we will but take the trouble to recall vividly the Mosaic cosmogony, in terms of which most of us have but recently ceased to frame our sublimest concep-tions of the vast: if we remind ourselves of Plato's " con-centric crystal spheres, the adamantine axis turning in the lap of necessity, the bands that held the heaven to-gether like a girth that clasps a ship, the shaft which led from earth to sky, and which was paced by the soul in a thousand years " ; if we compare these conceptions

[2] James Martineau: *Seat of Authority in Religion.*

with our own; if we think of " the fields from which our stars fling us their light," fields that are really near and yet are so far that the swiftest of messengers, capable of circling the earth eight times in a second, requires for its journey hither thousands of years; if we do but make some such comparisons, we shall begin to realize dimly that, compared with modern space — the space of modern thought — elder space — the space of elder thought — is indeed " but as a cabinet of brilliants, or rather a little jewelled cup found in the ocean or the wilderness."

Suppose that in fact space were thus, like time, not a constant, but a variable; suppose it were a mental thing growing with the growth of mind; an increasing function of increasing thought; suppose it were a thing whose enlargement is essential as a psychic condition or con-comitant or effect of the progress of science; would not our questions regarding its figure and its dimensions then lose their meaning? The answer is, no; as rational beings we should still be bound to ask: what are the dimensions and what is the figure of space to date? That is not all. If these questions were answered, we could propound the further questions: whether the space so characterized — the space of the present — is adequate to the present needs of science, and whether it is not destined to yet further expansion in response to the future needs of thought.

Men do not feel, however, that such spatial enlarge-ments as I have indicated are genuine enlargements of space. In spite of whatever metaphysics or psychology may seem obliged to say to the contrary, men feel that what is *new* in such an enlargement is merely an increase of enlightenment regarding something old; they feel that what is new is, not an added vastness, but a discovery of a vastness that always was and always will be. Let

us trust this feeling and, regarding space as constant from
everlasting to everlasting, let us take the questions in
their natural intent and form: what are the dimensions
and what is the figure of our universe of space?

If you propound these questions to a normal student
of natural science, say to a normal astronomer, his response will be — what? If you appear to him to be
quite sincere and if, besides, he be in an amiable mood,
his response will, not improbably, be a significant shrug
of the shoulders, designed to intimate that his time is
too precious to be squandered in considering questions
that, if not meaningless, are at all events unanswerable.[3]
I maintain, on the contrary, that this same student of
natural science, and indeed, all other normally educated
men and women, have, as a part of their intellectual
stock in trade, perfectly definite answers to both of the
questions. I do not mean that they are aware of pos-
sessing such wealth nor shall I undertake to say in
advance whether their answers be correct. What I am
asserting and what, with your assistance, I shall endeavor
to demonstrate, is that perfectly precise, very intelligent
and perfectly intelligible answers to both of the questions
are logically involved in what every normally educated
mind regards as the securest of its intellectual possessions.
In order to show that such answers are to be found em-
bedded in the content of the normally educated mind and
in order to lay them bare, it will be necessary to have
recourse to the process of explication. Explication, however, is nothing strange to an academic audience. It is
true, indeed, that we no longer derive the verb, to educate, from *educere,* but it is yet a fact, as every one
knows, that a large part of education is *eduction* — the

---

[3] Since this was written the theory of relativity has made astronomers and others more respectful in such matters.

leading forth into light what is hidden in the familiar content of our minds.

What are those answers? I shall present them in the familiar and brilliant words of one who in the span of a short life achieved a seven-fold immortality: immortality as a physicist, as a philosopher, as a mathematician, as a theologian, as a writer of prose, as an inventor and as a fanatic. From this brief but " immortal " characterization I have no doubt that you detect the author at once and at once recall the words: *Space is an infinite sphere whose center is everywhere and whose surface is nowhere.*

You will observe that, without change of meaning, I have substituted " space " for " universe " and " surface " for " circumference." This brilliant *mot* of Blaise Pascal, as every one knows, has long been valued throughout the world as a splendid literary gem. I am not aware that it has been at any time regarded seriously as a scientific thesis. It may, however, be so regarded. I propose to show, with your co-operation, that this exquisite saying of Pascal expresses with mathematical precision the firm, albeit unconscious, conviction of the normally educated mind respecting the size and shape of the space of our universe. Be good enough to note carefully at the outset the cardinal phrases: *infinite sphere, center everywhere, surface nowhere.*

If you are told that there is an object completely enclosed and that the object is equally distant from all parts of the enclosing boundary or wall, you instantly and rightly think of a sphere having that object as center. Let me ask you to think of some point, any convenient point, *P,* together with all the straight lines or rays — called a sheaf of lines or rays — that, beginning at *P,* run out from it as far as ever the nature of space allows. We ask: do all the rays of the sheaf run out equally far? It seems perfectly evident that they do, and with this we might be content. It will be worth while, however, to examine the matter a little more attentively. Denote by L any chosen line or ray of the sheaf. Choose any convenient unit of length, say a mile. We now ask: how many of our units, how many miles can we, starting from P, lay off along L? Lay off, I mean, not in fact, but in thought. In other words: how many steps, each a mile long, can we, in traversing L, take in thought? Hereafter let the phrase " in thought " be understood. Can the question be answered? It can. Can it be answered definitely? Absolutely so. How? As follows. Before proceeding, however, let me beg of you not to hesitate or shy if certain familiar ideas seem to get submitted to the logical process — the mind-expanding process — of generalization. There is to be no resort to any kind of legerdemain. Let us be willing to transcend imagination, and, without faltering, to follow thought, for thought, free as the spirit of creation, owns no bar save that of inconsistence or self-contradiction. Consider the sequence of cardinal numbers,

$$(S) \quad 1, \ 2, \ 3, \ 4, \ 5, \ 6, \ 7, \ \ldots \ldots$$

The sequence is neither so dry nor so harmless as it seems. It has a beginning; but it has no end, for, by the law of its formation, after each term there is a next. The difference between a sequence that stops somewhere and one that has no end is awful. No one, unless spiritually unborn or dead, can contemplate that gulf without emotions that take hold of the infinite and everlasting. Let us compare the sequence with the ray L of our sheaf. Choose in (S) any number n, however large. Can we go from P along L that number n of miles? We are certain that we can. Suppose the trip made, a mile post set up and on it painted the number n to tell how far the post is from P. As n is any number in (S), we may as well suppose, indeed we have already implicitly supposed, mile posts, duly distributed and marked, to have been set up along L to match each and every number in the sequence. Have we thus set up all the mile posts that L allows? We are certain that we have, for, if we go out from P along L any possible but definite number of miles, we are perfectly certain that that number is a number in the sequence, and that accordingly the journey did but take us to a post set up before. What is the upshot? It is that L admits of precisely as many mile posts as there are cardinal numbers, neither more nor less. How long is L? The answer is: L is exactly as many miles long as there are integers or terms in the sequence (S). Can we say of any other line or ray L' of the sheaf what we have said of L? We are certain that we can. Indeed we have said it, for L was *any* line of the sheaf. May we, then, say that any two lines, L and L', of the sheaf are *equal*? We may and we must. For, just as we have established a one-to-one correspondence between the mile posts of L and the terms of (S), so we may establish a one-to-one correspondence between the mile posts of L and those of L', and what we mean by the *equality* of two classes of things is precisely the possibility of thus setting up a one-to-one correlation between them. Accordingly, all the lines or rays of our sheaf are equal. We can not fail to note that thus there is forming in our minds the conception of a sphere, centered at P, larger, however, than any sphere of slate or wood or marble — a sphere, if it be a sphere, whose radii are the rays of our sheaf. Is not the thing, however, too vast to be a sphere? Obviously yes, if the lines or rays of the sheaf have a length that is indefinite,

unassignable; obviously no, if their length be assignable
and definite. We have seen the length of a ray contains
exactly as many miles as there are integers or terms in
(S). The question, then, is: has the totality of these
terms a definite assignable number? The answer is, yes.
To show it, look sharply at the following fact, a bit difficult to see only because it is so obvious, being writ, so
to speak, on the very surface of the eye. I wish, in a
word, to make clear what is meant by the cardinal number of any given class of things. The fingers of my right
hand constitute a class of objects; the fingers of my left
hand, another class. We can set up a one-to-one correspondence between the classes, pairing the objects in the
one with those in the other. Any two classes admitting
of such a correlation are said to be *equivalent*. Now
given any class K, there is another class C composed of
all the classes each of which is equivalent to K. C is
called the cardinal number of K, and the name of C, if
it has received a name, tells how many objects are in K.
Thus, if K is the class of the fingers of my right hand,
the word *five* is the name of the class of classes each
equivalent to K. Now to the application. The terms of
(S) constitute a class K (of terms). Has it a definite
number? Yes. What is it? It is the class of all classes
each equivalent to K. Has this number class received a
name of its own? Yes, and it has, like many other numbers, received a symbol, namely, $\aleph_0$, read Aleph null.
It is, then, this cardinal number Aleph, not familiar, indeed, but perfectly definite as denoting a definite class,
it is this that tells us how many terms are in (S) and
therewith tells us the length of the rays of our sheaf.
Herewith the concept that was forming is now completely
formed: *space is a sphere centered at P.*

But is the sphere, as Pascal asserts, an *infinite* sphere?

We may easily see that it is. Again consider the sequence
(S) and with it the similar sequence (S'),

$$(S) \quad 1, 2, 3, 4, 5, 6, 7, \ldots,$$
$$(S') \quad 2, 4, 6, 8, 10, 12, 14, \ldots\ldots$$

Observe that all the terms in (S') are in (S) and that
(S) contains terms that are not in (S'). (S') is, then,
a proper *part* of (S). Next observe that we can pair
each term in (S) with the term below it in (S'). That
is to say: the whole, (S), is equivalent to one of its parts,
(S'). A class that thus has a part to which it is equivalent is said to be infinite,[4] and the number of things in
such a class is called an infinite number. Aleph is, then,
an infinite number, and so we see that the rays of our
sheaf, the radii of our sphere, are infinite in length:
space is an infinite sphere centered at P.

Finally, what of the phrases, *center everywhere, surface
nowhere?* Can we give them a meaning consistent with
common usage and common sense? We can, as follows.
Let O be any chosen point somewhere in your neighbor-
hood. By saying that the center P is everywhere we
mean that P may be taken to be *any* point within a
sphere centered at O and having a finite radius, a radius,
that is, whose length in miles is expressed by any integer
in (S). And by saying that the surface of our infinite
sphere is nowhere we mean that no point of the surface
can be reached by traveling out from P any *finite* number, however large, of miles, by traveling, that is a
number of miles expressed by any number, however
large, in (S).

Here we have touched our primary goal: we have
demonstrated that men and women whose education, in

---

[4] For an exposition of the modern concept of infinity with its bearings in philosophy see Keyser's *The New Infinite and the Old Theology,*
The Yale University Press.

respect of space, has been of normal type, believe profoundly, albeit unawares, that the space of our universe is an infinite sphere of which the center is everywhere and the surface nowhere. Such is the beautiful conception, the great conception — mathematically precise yet mystical withal and awful in its limitless reaches — which is ever ready to form itself, in the normally educated mind and there to stand a deep-rooted conscious conviction regarding the shape and the size of the all-embracing world.

Is the conception valid? Does the conviction correspond to fact? Is it true? It is not enough that it be intelligible, which it is; it is not enough that it be noble and sublime, which also it is. No doubt whatever is noble and sublime is, in some sense, true. For we mortals have to do with more than reason. Yet science, science in the modern technical sense of the term, having elected for its field the domain of the rational, allows no superrational tests of truth to be sufficient or final. We must, therefore, ask: are the dimensions and the figure of our space, in fact, what, as we have seen, Pascal asserts and the normally educated mind believes them to be? Long before the days of Pascal, back yonder in the last century before the beginning of the Christian era, one of the acutest and boldest thinkers of all time, immortal expounder of Epicurean thought, answered the question, with the utmost confidence, in the affirmative. I refer to Lucretius and his " De Rerum Natura." In my view that poem is the greatest and finest union of literary excellence and scientific spirit to be found in the annals of human thinking. I maintain that opinion of the work despite the fact that the majority of its conclusions have been invalidated by time, have perished by supersession; for we must not forget that, in respect of knowledge, " the present is no more exempt from the sneer of the future than the past has been." I maintain that opinion of the work despite the fact that the enterprise of Lucretius was marvelously extravagant; for we must not forget that the relative modesty of modern men of science is not inborn, but is only an imperfectly acquired lesson. Well, it is in that great work that Lucretius endeavors to prove that our universe of space is infinite in the sense that we have explained. His argument, which runs to many words, may be briefly paraphrased as follows. Conceive that, starting from any point of space, you go out in any direction as far as you please, and that then you hurl your javelin. Either it will go on, in which case there is space ahead for it to move in, or it will not go on, in which case there must be space ahead to contain whatever prevents its going. In either case, then, however far you may have gone, there is yet space beyond. And so, he concludes, space is infinite, and he triumphantly adds:

Therefore the nature of room and the space of the unfathomable void are such as bright thunderbolts can not race through in their course though gliding on through endless tract of time, no nor lessen one jot the journey that remains to go by all their travel — so huge a room is spread out on all sides for things without any bounds in all directions round.

Such is the argument, the great argument, of the Roman poet.[5] Great I call it, for it is great enough to have fooled all philosophers and men of science for two thousand years. Indeed only a decade ago I heard the argument confidently employed by an American thinker of more than national reputation. But is the argument

[5] As to the rôle of *infinity* in the great work of Lucretius see the chapter on Infinity in Keyser's *Mathematical Philosophy*, Dutton & Company.

really fallacious? It is. The conclusion may indeed be quite correct — space may indeed be infinite, as Lucretius asserts — but it does not follow from his argument. To show the fallacy is no difficult feat. Consider a sphere of finite radius. We may suppose it to be very small or intermediate or very large — no matter what its size so long as its radius is finite. By sphere, in this part of the discussion, I shall mean sphere-surface. Be good enough to note and bear that in mind. Observe that this sphere — this surface — is a kind of room. It is a kind of space, region or room where certain things, as points, circle arcs and countless other configurations can be and move. These things, confined to this surface, which is their world, their universe of space, if you please, enjoy a certain amount, an immense amount, of freedom: the points of this world can move in it hither, thither and yonder; they can move very far, millions and millions of miles, even in the same direction, if only the sphere be taken large enough. I see no reason why we should not, for the sake of vividness, fancy that spherical world in- habited by two-dimensional intelligences conformed to their locus and home just as we are conformed to our own space of three dimensions. I see no reason why we should not fancy those creatures, in the course of their history, to have had their own Democritus and Epicurus, to have had their own Roman republic or empire and in it to have produced the brilliant analogues of our own Virgil, Cicero and Lucretius. Do but note attentively — for this is the point — that their Lucretius could have said about their space precisely what our own said about ours. Their Lucretius could have said to his fellow- inhabitants of the sphere: " starting at any point, go as far as ever you please in any straight line " — such line would of course (as *we* know) be a great circle of the

sphere — " and then hurl your javelin " — the javelin
would, as *we* know, be only an *arc* of a great circle —
" either it will go on, in which case, etc.; or it will not,
etc." ; thus giving an argument identical with that of our
own Lucretius. But what could it avail? *We* know what
would happen to the javelin when hurled as supposed in
the surface: it would go on for a while, there being nothing to prevent it. But whether it went on or not, it
could not be logically inferred that the surface, the space
in question, is infinite, for we know that the surface is
finite, just so many, a finite number of, square miles.
The fallacy, at length, is bare. It consists — the fact
has been recently often pointed out — in the age-long
failure to distinguish adequately between *unbegrenzt* and
unendlich — between *boundless* and *infinite* as applied
to space. What our fancied Lucretius proved is, if anything, that the sphere is boundless, but not that it is
infinite. What our real Lucretius proved is, if anything,
that the space of our universe is boundless, but not that
it is infinite. That a region or room may be boundless
without being infinite is clearly shown by the sphere
(surface). How evident, once it is drawn, the distinction
is. And yet it was never drawn, in thinking about the
dimensions of space, until in 1854 it was drawn by Riemann in his epoch-marking and epoch-making *Habilitationschrift* on the foundations of geometry.

What, then, is the fact? Is space finite, as Riemann
held it may be? Or is it infinite, as Lucretius and Pascal
deliberately asserted, and as the normally educated mind,
however unconsciously, yet firmly believes? No one
knows. The question is one of the few great outstanding
scientific questions that intelligent laymen may, with a
little expert assistance, contrive to grasp. Shall we ever
find the answer? Time is long, and neither science nor

philosophy feels constrained to haul down the flag and confess an *ignorabimus*. Neither is it necessary or wise for science and philosophy to camp indefinitely before a problem that they are evidently not yet equipped to solve. They may proceed to related problems, always reserving the right to return with better instruments and added light.

In the present instance, let us suppose, for the moment, that Lucretius, Pascal and the normally educated mind are right: let us suppose that space is infinite, as they assert and believe. In that case the bounds of the universe are indeed remote, and yet we may ask: are there not ways to pass in thought the walls of even so vast a world? There are such ways. But where and how? For are we not supposing that the walls to be passed are distant by an amount that is infinite? And how may a boundary that is infinitely removed be reached and overpassed? The answer is that there are many infinites of many orders; that infinites are surpassed by other infinites; that infinites, like the stars, differ in glory. This is not rhetoric, it is naked fact. One of the grand achievements of mathematics in the nineteenth century is to have defined infinitude (as above defined) and to have discovered that infinites rise above infinites, in a genuine hierarchy without a summit. In order to show how we can in thought pass the Lucretian and Pascal walls of our universe, I must ask you to assume as a lemma a mathematical proposition which has indeed been rigorously established and is familiar, but the proof of which we can not tarry here to reproduce. Consider all the real numbers from *zero* to *one* inclusive, or, what is tantamount, consider all the points in a unit segment of a continuous straight line. The familiar proposition that I am asking you to assume is that it is not possible to set up a one-to-one correspondence between the points of that segment and the positive integers (in the sequence above given), but that, if you take away from the segment an infinitude (Aleph) of points matching all the positive integers, there will remain in the segment more points, infinitely more, than you have taken away. That means that the infinitude of points in the segment infinitely surpasses the infinitude of positive integers; surpasses, that is, the infinitude of mile posts on the radius of our infinite (Pascal) sphere. Now conceive a straight line containing as many miles as there are points in the segment. You see at once that in that conception you have overleaped the infinitely distant walls of the Lucretian universe. Overleaped, did I say? Nay, you have passed beyond those borders by a distance infinitely greater than the length of any line contained within them. And thus it appears that, not our imagination, indeed, but our reason may gaze into spatial abysses beside which the infinite space of Lucretius and Pascal is but a meagre thing, infinitesimally small. There remain yet other ways by which we are able to escape the infinite confines of this latter space. One of these ways is provided in the conception of hyperspaces enclosing our own as this encloses a plane. But that is another story, and the hour is spent.

The course we have here pursued has not, indeed, enabled us to answer with final assurance the two questions with which we set out. I hope we have seen along the way something of the possibilities involved. I hope we have gained some insight into the meaning of the questions and have seen that it is possible to discuss them profitably. And especially I hope that we have seen afresh, what we have always to be learning again, that it is not in the world of sense, however precious it is and ineffably wonderful and beautiful, nor yet in the still finer and ampler world of imagination, but it is in the world of conception and thought that the human intellect attains its appropriate freedom — a freedom without any limitation save the necessity of being consistent. Consistency, however, is only a prosaic name for a limitation which, in another and higher realm, harmony imposes even upon the muses.

MATHEMATICAL EMANCIPATIONS: DIMENSIONALITY AND HYPERSPACE [1]

AMONG the splendid generalizations effected by modern mathematics, there is none more brilliant or more inspiring or more fruitful, and none more nearly commensurate with the limitless immensity of being itself, than that which produced the great concept variously designated by such equivalent terms as hyperspace, multidimensional space, n-space, n-fold or n-dimensional space, and space of n dimensions.

In science as in life the greatest truths are the simplest. Intelligibility is alike the first and the last demand of the understanding. Naturally, therefore, those scientific generalizations that have been accounted really great, such as the Newtonian law of gravitation, or the principle of the conservation of energy, or the all-conquering concept of cosmic evolution, are, all of them, distinguished by their simplicity and apprehensibility. To that rule the notion of hyperspace presents no exception. For its fair understanding, for a live sensibility to its manifold significance and quickening power, a long and severe mathematical apprenticeship, however helpful it would be, is not demanded in preparation, but only the serious attention of a mature intelligence reasonably inured by discipline to the exactions of abstract thought and the austerities of the higher imagination. And it is to the reader having this general equipment, rather than

---

[1] Printed in *The Monist*, January, 1906. For a deeper view of this subject the reader may be referred to the 13th essay of this volume.

to the professional mathematician as such, that the present communication is addressed.

To a clear understanding of what the mathematician means by hyperspace, it is in the first place necessary to conceive in its full generality the closely related notion of dimensionality and to be able to state precisely what is meant by saying that a given manifold has such and such a dimensionality, or such and such a number of dimensions, in a specified entity or element.

Discrimination, as the proverb rightly teaches, is the beginning of mind. The first psychic product of that initial psychic act is *numerical:* to discriminate is to produce *two,* the simplest possible example of multiplicity. The discovery, or better the invention, better still the production, best of all the creation, of multiplicity with its correlate of number, is, therefore, the most primitive achievement or manifestation of mind. Such creation is the immediate issue of intellection, nay, it *is* intellection, identical with its deed, and, without the possibility of the latter, the former itself were quite impossible. Accordingly it is not matter for surprise but is on the contrary a perfectly natural or even inevitable phenomenon that explanations of ultimate ideas and ultimate explanations in general should more and more avail themselves of analytic as distinguished from intuitional means and should tend more and more to assume arithmetic form. Depend upon it, the universe will never really be understood unless it may be sometime resolved into an ordered multiplicity and made to own itself an everlasting drama of the calculus.

Let us, then, trust the arithmetic instinct as fundamental and, for instruments of thought that shall not fail, repair at once to the domain of number. Every one who may reasonably aspire to a competent knowledge of

the subject in hand is more or less familiar with the
system of real numbers, composed of the positive and
negative integers and fractions, such irrational numbers
as $\sqrt{2}$ and π and countless hosts of similar numbers
similarly definable. He may know that, for reasons
which need not be given here, the system of real numbers
is commonly described as the analytical continuum of
second order. He knows, too, at any rate it is a fact
which he will assume and readily appreciate, that the
distance between any two points of a right line is exactly
expressible by a number of the continuum; that, conversely, given any number, two points may be found
whose distance apart is expressed by the numerical value
of that number; that, therefore, it is possible to establish
a unique and reciprocal, or one-to-one, correspondence
between the real numbers and the points of a straight
line, namely, by assuming some point of the line as a
fixed point of reference or origin of distances, by agreeing that a distance shall be positive or negative according as it proceeds from the origin in this sense or in the
other and by agreeing that a *point* and the *number* which
by its magnitude reckoned in terms of a chosen finite
unit however great or small serves to express the distance of the point from the origin and by its sign indicates on which side of the origin the point is situated,
shall be a pair of *correspondents*. Accordingly, if the
point P glides along the line, the corresponding number
v will vary in such a way that to each position of the
geometric there corresponds one value of the arithmetic
element, and conversely. P represents v; and v, P. No
two P's represent a same v; and no two v's, a same P.
By virtue of the correlation thus established with the
analytical continuum, we may describe the line as a
simple or one-fold *geometric* continuum, namely, of

points. The like may in general be said, and for the same reason, of any curve whatever, but we select the straight line as being the simplest, for in matters fundamental we should prefer clearness to riches of illustration, in the faith that, if first we seek the former, the latter shall in due course be added unto it. The straight line, when it is regarded as the domain of geometric operation, as the region or room containing the configurations or sets of elements with which we deal, is and is called a *space;* and this space, viewed as the manifold or assemblage of its points, is said to be *one*-dimensional for the reason that, as we have seen, in order to determine the position of a point in it, in order, *i.e.*, to pick out or distinguish a point from all the other points of the manifold, it is necessary and sufficient to know one fact about the point, as *e.g.*, its distance from an assumed point of reference. In other words, the line is called a one-dimensional space of points because in that space the point has one and but one degree of freedom or, what is tantamount, because the position of the point depends upon the value of a single v, known as its coordinate.

Herewith is immediately suggested the generic concept of dimensionality: *if an assemblage of elements of any given kind whatsoever, geometric or analytic or neither, as points, lines, circles, triangles, numbers, notions, sentiments, hues, tones, be such that, in order to distinguish every element of the assemblage from all the others, it is necessary and sufficient to know exactly n independent facts about the element, then the assemblage is said to be n-dimensional in the elements of the given kind.* It appears, therefore, that the notion of dimensionality is by no means exclusively associated with that of space but on the contrary may often be attached to the far more generic concept of assemblage, aggregate or manifold. For example, duration, the total aggregate of time-points, or instants, is a simple or one-fold assemblage. On the other hand, the assemblage of colors is three-dimensional as is also that of musical notes, for in the former case, as shown by Clerk Maxwell, Thomas Young and others, every color is composable as a definite mixture of three primary ones and so depends upon three independent variables or coordinates expressing the amounts of the fundamental components. And in the latter case a similar scheme obtains, one note being distinguishable from all others when and only when the three general marks, pitch, length, and loudness, are each of them specified. In passing it seems worth while to point out the possibility of appropriating the name *soul* to signify the *manifold* of all *possible* psychic *experiences,* in which event the term would signify an assemblage of probably *infinite* dimensionality, and the assemblage would be continuous, too, if Oswald [2] be right in his contention that every manifold of experience possesses the character of continuity. That contention, however, if the much abused term continuity be allowed to have its single precise definitely seizable scientific meaning, is far less easy to make good than that eminent chemist and courageous philosopher seems to think.

Returning to the concept of space, an *n*-fold *assemblage* will be an *n*-dimensional space if the elements of the assemblage are geometric entities of any given kind. We have seen that the straight line is a *one*-dimensional space of *points*. But in studying the right line conceived as a space, we are not compelled to employ the point as element. Instead we may choose to assume as element the point *pair* or *triplet* or *quatrain,* and so on The line

2 *Cf.* his *Natur-Philosophie.*

would then be for our thought primarily a space, not of points, but of point pairs or triplets and so on, and it would accordingly be strictly a space of *two dimensions* or of *three*, and so on; for, obviously, to distinguish say a point pair from all other such pairs we should have to know *two* independent facts about the pair. The pair would have two degrees of freedom in the line, its determination would depend upon two independent variables as v_1 and v_2. These variables might be the two independent ratios of the coefficients and absolute term in a quadratic equation in one unknown, as x, for to know the ratios is to know the equation and therewith its two roots, the two values of x. These being laid off on the line give the point pair. Conversely, a point pair gives two values of x, hence a definite quadratic equation and so values of v_1 and v_2. On its arithmetic side the shield presents a precisely parallel doctrine. The simple analytical continuum composed of the real numbers immediately loses its simplicity and assumes the character of a 2- or 3- . . . or n-fold analytical continuum if, instead of thinking of its individual numbers, we view it as an aggregate of number pairs or triplets or, in general, as the totality of ordered systems of n numbers each.

In the light of the preceding paragraph it is seen that the dimensionality of a given space is not unique but depends upon the choice of geometric entity for primary or generating element. A space being given, its dimensionality is not therewith determined but depends upon the will of the investigator, who by a proper choice of generating elements may endow the space with any dimensionality he pleases. That fact is of cardinal significance alike for science and for philosophy. I reserve for a little while its further consideration in order to present at once a kind of complementary fact of equal interest and of scarcely less importance. It is that two spaces which in every other respect are essentially unlike, thoroughly disparate, may, by suitable choice of generating elements, be made to assume equal dimensionalities. Consider, for example, the totality of lines contained in a same plane and containing a point in common. Such a totality, called a *pencil*, of lines is a simple geometric continuum, namely, of lines. It is, then, and may be called, a *one*-dimensional space of *lines* just as the line or *range* of points is a one-dimensional space of points. The two spaces are equally rich in their respective elements. And if, following Desargues and his successors, we adjoin to the points of the range a so-called " ideal " point or point at infinity, thus rendering the range like the pencil, *closed,* it becomes obvious that two intelligences, adapted and confined respectively to the two simple spaces in question, would enjoy equal freedom; their analytical experiences would be identical, and their geometries, though absolutely disparate in kind, would be equally rich in content. Just as the range-dweller would discover that the dimensionality of his space is *two* in *point pairs, three* in *triplets,* and so on, so the pencil-inhabitant would find his space to be of dimensionality *two* in *line pairs, three* in *triplets,* and so on without end. It was indicated above that *any* line, straight or *curved,* is a *one*-dimensional space of *points.* In that connection it remains to say that, speaking generally, *any curve,* literally and strictly conceived as the assemblage of its (tangent) lines and so including the point or pencil as a special case, is also a *one*-dimensional space of *lines.* It is, moreover, obvious that the foregoing considerations respecting the range of points and the pencil of lines are, *mutatis mutandis,* equally valid for any one of an infinite variety of other analogous spaces, as, *e.g.,* the *axial pencil,* a one-fold space of planes, consisting of the totality of planes having a line in common.

If perchance some reader should feel an ungrateful sense of impropriety in our use of the term *space* to signify such common geometric aggregates as we have been considering, I gladly own that his state of mind is a perfectly natural one. But it is, besides and on that account, a source of real encouragement. Dictional sensibility is a hopeful sign, being conclusive evidence of life, and, while there is life, there remains the possibility and therewith the hope of readjustment. In the present case, I venture to assure the reader, on grounds both of personal experience and of the experience of others, that whatever sense he may have of injury received will speedily disappear in the further course of his meditations. Only, let him not be impatient. Larger meanings must have time to grow; the smaller ones, those that are most natural and most provincial, being also the most persistent. In the process of clarification, expansion and readjustment, his fine old word, space, early come into his life and gradually stained through and through with the refracted partial lights and multi-colored prejudices of his youth, is not to be robbed of its proper charms nor to be shorn of its proper significance. More than it will lose of mystery, it shall gain of meaning. Of this last it has hitherto had for him but little that was of scientific value, but little that was not vague and elusive and ultimately unseizable. That was because the word stood for something absolutely *sui generis, i. e., for a genus neither including species nor being itself included in a class.* But now, on the other hand, both of these negatives are henceforth to be denied, and the hitherto baffling term, perfect symbol of the unthinkable, always promising and never presenting definable content, immediately assumes the characteristic twofold aspect of a genuine concept, being at once included as member of a higher class, the more generic class of manifolds, and including within itself an endless variety of individuals, an infinitude of species of space.

Of these species, the next in order of simplicity, to those above considered, is the plane. To distinguish a point of a plane from all its other points, it is necessary and sufficient to know *two* independent facts about its position, as, *e.g.*, its distances from two assumed lines of reference, most conveniently taken at right angles. Viewed as the *ensemble* of its points, the plane is, therefore, a space of *two* dimensions. In that space, the point enjoys a freedom exactly twice that of a point in a range or of a line in a pencil, and exactly equal to that of a *pair* of points or of lines in the last-mentioned spaces. On the other hand, if the point *pair* be taken as element of the plane, the latter becomes a space of *four* dimensions.

What if the *line* be taken as generating element of the plane? It is obvious that the plane is equally rich in pencils and in ranges. It contains as many lines as points, neither more nor less. Two points determine a line; two lines, a point; if the lines be parallel, their common point is a Desarguesian, a point at infinity. We should therefore expect to find that in a plane the position of a line depends upon two and but two independent variables. And the expectation is realized, as it is easy to see. For if the variables be taken to represent (say) distances measured from chosen points along two lines of reference, it is immediately evident that a given pair of values of the variables determines a line uniquely and that, conversely, a given line uniquely determines such a pair. The plane is, therefore, a *two*-dimensional space of *lines* as well as of points. In line *pairs*, as in point pairs, its dimensionality is *four*. We may suppose the space in question to be inhabited by two sorts of individuals, one of them capable of thinking in terms of points but not of lines, the other in terms of lines but not of points. Each would find his space bi-dimensional. They would enjoy precisely the same analytical experience. Between their geometries there would subsist a fact-to-fact *correspondence* but not the slightest *resemblance*. For example, the circle would be for the former individual a certain assemblage of points but devoid of tangent lines, and, for the latter, a corresponding assemblage of (tangent) lines but devoid of contact points.

Passing from the plane to a curved surface, to a sphere, for example, a little reflection suffices to show that the latter may be conceived in a thousand and one ways, but most simply as the ensemble of its points *or* of its (tangent) planes *or* of its (tangent) lines. These various concepts are logically equivalent and in themselves are equally intelligible. And if to us they do not seem to be also equally *good*, that is doubtless because we are but little traveled in the great domain of Reason and therefore naturally prefer our familiar customs and provincial points of view to others that are strange. At all events, it is certain that on purely *rational* grounds, none of the concepts in question is to be preferred, while, from preference based on *other* grounds, it is the office alike of science and of philosophy to provide the means of emancipation. Let us, then, detach ourselves from the vulgar point of view and for a moment contemplate the three concepts as coördinate indeed but independen't concepts of surface. And for the sake of simplicity, we may think of a sphere.[3] Suppose it placed upon a plane and imagine its highest point, which we may call the

---

[3] The term is here employed as in the higher geometry to denote, not a solid, but a surface.

pole, joined by straight lines to all the points of the plane. Each line pierces the sphere in a second point. It is plain that thus a one-to-one correspondence is set up between the points of the sphere and those of the plane, except that the pole corresponds at once to all the Desarguesian points of the plane — an exception, however, which is here of no importance. The plane and the sphere are, then, equally rich in points. Accordingly, the sphere conceived as a plenum or locus or space of *points* is a space of *two* dimensions. In that space the point has two degrees of freedom. Its position depends upon two independent variables, as latitude and longitude. But we may conceive the surface quite otherwise: at each of its points there is a (tangent) plane, and now, *disregarding points*, we may think only of the assemblage of those planes. These together constitute a sphere, not, however, as a locus of points, but as an envelope (as it is called) of planes. And what shall we say of the surface as thus conceived? The answer obviously is that it is a *two*-dimensional space of *planes*, admitting of a geometry quite as rich and as definite as is the theory of any other space of equal dimensionality. In each of the planes there is a pencil of lines of which each is tangent to the sphere. Thus we are led to a third conception of our surface. We have merely to disregard both points and planes and confine our attention to the assemblage of lines. The vision which thus arises is that of a *three*-dimensional space of *lines*. In pencils, its dimensionality is two. In this space the pencil has two and the line three degrees of freedom.

But let us return to the plane. We have seen that at the geometrician's bidding it plays the rôle of a two-fold space either in points or in lines. It is natural to ask whether it may be conceived as a space of *three* dimensions, like the sphere in its third conception. The answer is affirmative: it may be so conceived, and that in an infinity of ways. Of these the simplest is to assume the *circle* as primary or generating element. Of circles the plane contains a threefold infinity, an infinity of infinities of infinities. It is a circle continuum of third order. To distinguish any one of its circles from all the rest, three independent data, two for position and one for size, are necessary and sufficient. In the plane the circle has three degrees of freedom, its determination depends upon three independent variables. The plane is, accordingly, a tri-dimensional space of *circles*. In *parabolas* its dimensionality is *four;* in *conics, five;* and so on *without limit.*

Before turning to space, ordinarily so-called, it seems worth while to indicate another geometric continuum which, although it presents no *likeness* whatever to the plane, nevertheless *matches* it perfectly in every conceptual aspect. The reference is to the *sheaf,* or *bundle,* of *lines, i.e.,* the totality of lines having a point in common. The point is to be disregarded and the lines viewed as non-decomposable entities, like points in a line or plane regarded as an assemblage of points. Thus conceived, the *sheaf* is literally a *space,* namely, of *lines.* It is, in the vulgar sense of the term, just as *big,* occupies precisely as much room, nay indeed the same room, as the space in which we live. The sheaf as a space is *two*-dimensional in *lines,* like the *plane* in *points; two*-dimensional in *pencils,* like the *plane* in *lines; four*-dimensional in *line* or *pencil pairs,* like the *plane* in *point* or *line pairs; three*-dimensional in ordinary *cones,* like the *plane* in *circles;* and so on and on.

In the light of the foregoing considerations, any hitherto uninitiated reader will probably suspect that ordinary space is *not,* as it is commonly supposed and said to be, an *inherently* and *uniquely three-*dimensional affair. His suspicion is completely justified by fact. The simple traditional affirmation of tri-dimensionality is devoid of definite meaning. It is unconsciously elliptic, requiring for its completion and precision the specification of an appropriate geometric entity for generating element. Merely to say that space is tri-dimensional because a solid, *e.g.,* a plank, has length, breadth and thickness, is too crude for scientific purposes. Moreover, it betrays, quite unwittingly indeed as we shall see, an exceedingly meager point of view. Not only does it assume the *point* as element but it does so tacitly because unconsciously, as if the point were not merely *an* but *the* element of ordinary space. *An* element the point may obviously be taken to be, and in that element ordinary space is indeed tri-dimensional, for the position of a point at once determines and is determined by three independent data, as its distances from three assumed mutually perpendicular planes of reference. It must be admitted, too, that the point does, in a sense, recommend itself as the element *par excellence,* at least for practical purposes. For example, we prefer to do our drawing with the point of a pencil to doing it with a straight edge. But that is a matter of physical as distinguished from rational convenience. Preference for the point has, then, a cause: in the order of evolution, practical man precedes man rational and determines for the latter his initial choices. Causes, however, are extra-logical things, and the preference in question, though it has indeed a cause, has no reason. Accordingly, when in these modern times, the geometrician became clearly conscious that he was in fact and had been from time immemorial employing the point as element and that it was this use that lent to space its traditional triplicity of dimensions, he did not fail to perceive almost immediately the logically equal possibility of adopting at will for primary element any one of an infinite variety of other geometric entities and so the possibility of rationally endowing ordinary space with any prescribed dimensionality whatever.

Thus, for example, the plane is no less available for generating element than is the point. The plane is logically and intuitionally just as simple, for, if from force of habit, we are tempted to analyze the plane into an assemblage of points, the point is in its turn equally conceivable as or analyzable into an assemblage of planes, the sheaf of planes containing the point. We may, then, regard our space as primarily a plenum of planes. To determine a plane requires three and but three independent data, as, say, the distances to it measured along three chosen lines from chosen points upon them. It follows that ordinary space is *three*-dimensional in *planes* as well as in points. But now if (with Plücker) we think of the *line* as element, we shall find that our space has *four* dimensions. The fact may be seen in various ways, most easily perhaps as follows. A line is determined by any two of its points. Every line pierces every plane. By joining the points of one plane to all the points of another, all the lines of space are obtained. To determine a line it is, then, enough to determine two of its points, one in the one plane and one in the other. For each of these determinations, two data, as before explained, are necessary and sufficient. The position of the line is thus seen to depend upon four independent variables, and the four-dimensionality of our space in lines is obvious. Again, we may (with Lie) view our space as an assemblage of its *spheres*. To distinguish a sphere from all other spheres, we need to know four and but four independent facts about it, as, say, three that shall determine its center and one its size. Hence our space is *four*-dimensional also in *spheres*. In *circles* its dimensionality is *six;* in *surfaces* of *second order* (those that are pierced by a straight line in two points), *nine;* and so on *ad infinitum.*

Doubtless the reader is prepared to say that, if the foregoing account of hyperspace be correct, the notion is after all a very simple one. Let him be assured, the account is correct and his judgment is just: the notion is simple. That property, as said in the beginning, is indeed one of its merits. As presented the concept is entirely free from mystery. To seize upon it, it is unnecessary to pass the bounds of the visible universe or to transcend the limits of intuition. Its realization is found even in the line, in the pencil, in the plane, in the sheaf, here, there and yonder, everywhere, in fact. The account, however, though quite correct, is not yet complete. The term hyperspace has yet another meaning and yet in strictness not *another,* as we shall see. It will be noticed that among the foregoing examples of hyperspace, none is presented of dimensionality *exceeding three* in *points.* It is precisely this variety of hyperspace that the term is commonly employed to signify, particularly in popular enquiry and philosophical speculation. And it is this variety, too, that just because it baffles the ordinary visual imagination, proves to be, for the non-mathematician at any rate, at once so tantalizing, so mysterious and so fascinating.

It remains, then, to ask, what is meant by a hyperspace of *points?* How is the notion formed and what is its motivity and use? The path of enquiry is a familiar one and is free from logical difficulty. Granted that a one-to-one correspondence can be established between the real numbers and the points of a right line, so that the geometric serve to represent the arithmetic elements; granted that all (ordered) pairs of numbers are similarly representable by the points of a plane, and all (ordered) triplets by the points of ordinary space; the suggestion then naturally presents itself that, whether there really is or not, there *ought* to be a *space* whose *points* would serve to *represent,* as in the preceding cases, *all ordered systems* of *values* of *n independent variables;* and especially to an analyst with a strong geometric pre-dilection, to one who is a born *Vorstellender* for whom analytic abstractions naturally tend to take on figure and assume the exterior forms of sense, that suggestion comes with a force which he alone perhaps can fully appreciate. And what does he do? Not finding the desiderated hyperspace present to his *vision* or *intuition* or *visual imagination,* he *posits* it, or if you prefer, he *creates* it, *in thought.* The concept of hyperspace of points is thus seen to be offspring of Arithmetic and Geometry. It is legitimate fruit of the indissoluble union of the funda-mental sciences.

Does such hyperspace exist?   It does exist genuinely. If not for intuition, it exists for conception; if not for imagination, it exists for thought; if not for sense, it exists for reason; if not for matter, it exists for mind. These if's are *if's* in fact. The question of imaginability is really a question. We shall return to it presently.

The concept of hyperspace of points is generable in various other ways. Of all ways the following is perhaps the best because of its appeal at every stage to intuition. Let there be two points and grant that these determine a *line,* point-space of *one* dimension. Next posit a point *outside* of this line and suppose it joined by lines to all the points of the given line. The points of the joining lines together constitute a *plane*, point-space of *two* dimensions. Next posit a point outside of this plane and suppose it joined by lines to all the points of the plane. The points of all the joining lines together constitute an ordinary *space*, point-space of *three* dimensions. The clue being now familiar to our hand, let us boldly pursue the opened course. Let us overleap the limits of common imagination, transcend ordinary intuition as being at best but a non-essential auxiliary, and in *thought* posit an extra point that, *for thought* at all events, shall be *outside* the space last generated. Suppose that point joined by lines to all the points of the given space. The points of the joining lines together constitute a point-space of *four* dimensions. The process here applied is perfectly clear and obviously admits of endless repetition.

Moreover, the process is equally available for generating hyperspaces of other elements than points. For example, let there be two intersecting lines and grant that these determine a *pencil*, line-space of *one* dimension. Next posit a line (through the vertex) *outside* of the given pencil and suppose it joined by pencils to all the lines of the given pencil. The lines of the joining pencils together constitute a *sheaf*, line-space of *two* dimensions. Next posit a line (through the vertex) *outside* of the sheaf and suppose it joined by pencils to all the lines of the sheaf. The lines of the joining pencils constitute a *hypersheaf*, line-space of *three* dimensions. The next step plainly leads to a line-space of *four* dimensions; and so on *ad infinitum*.

And now as to the question of imaginability. Is it possible to intuit configurations in a hyperspace of points? Let it be understood at the outset that that is not in any sense a mathematical question, and mathematics as such is quite indifferent to whatever answer it may finally receive. Neither is the question primarily a question of philosophy. It is first of all a psychological question. Mathematicians, however, and philosophers are also men and they may claim an equal interest perhaps with others in the profounder questions concerning the potentialities of our common humanity. The question, as stated, undoubtedly admits of affirmative answer. For the lower spaces, with which the imagination *is* familiar, exist *in* the higher, as the line in the plane, and the plane in ordinary space. But that is not what the question means. It means to ask whether it is possible to imagine hyperconfigurations of points, *i.e.*, point-configurations that are not wholly contained in a point-space (like our own) of three dimensions. It is impossible to answer with absolute confidence. One reason is that the term imagination still awaits precision of definition. Undoubtedly just as three-dimensional figures may be represented in a plane, so four-dimensional figures may be represented in space. That, however, is hardly what is meant by imagining them. On the other hand, a four-dimensional figure may be rotated and translated in such a way that all of its parts come one after another into the threefold domain of the ordinary intuition. Again, *the structure of a fourfold figure, every minutest detail of its anatomy, can be traced out by analogy with its three-dimensional analogue.* Now in such processes, repetition yields skill, and so they come ultimately to require only amounts of *energy* and of *time* that are quite inappreciable. Such skill once attained, the *parts* of a familiar *fourfold configuration* may be made to pass before the eye of intuition in such *swift* and *effortless* succession that the configuration *seems present as a whole in a single instant.* If the *process* and *result* are not, properly speaking, *fourfold imagination* and *fourfold image,* it remains for the psychologist to indicate what is lacking.

Certainly there is naught of absurdity in supposing that

under suitable stimulation the human mind may in course of time even speedily develop a spatial intuition of four or more dimensions. At present, as the psychologists inform us and as every teacher of geometry discovers independently, the spatial imagination, in case of very many persons, comes distinctly short of being strictly even tri-dimensional. On the contrary, it is flat. It is not every one, even among scholars, that with eyes closed can readily form a visual image of the *whole* of a simple *solid* like a sphere, enveloping it completely with bent beholding rays of psychic light. In such defect of imagination, however, there is nothing to astonish. In the first place, man as a race is only a child. He has been on the globe but a little while, long indeed compared with the fleeting evanescents that constitute the most of common life, but very short, the merest fraction of a second, in the infinite stretch of time. In the second place, circumstances have not, in general, favored the development of his higher potentialities. His chief occupation has been the destruction and evasion of his enemies, contention for mere existence against hostile environment. Painful necessity, then, has been the mother of his inventions. That, and not the vitalizing joy of self-realization, has for the most part determined the selection of the fashion of his faculties. But it would be foolish to believe that these have assumed their final form or attained the limits of their potential development. The imperious rule of necessity will relax. It will never pass quite away but it will relax. It is relaxing. It has relaxed appreciably. The intellect of man will be correspondingly quickened. More and more will joy in its activity determine its modes and forms. The multidimensional worlds that man's reason has already created, his imagination may yet be able to depict and illuminate.

It remains to ask, finally, what purpose the concept of hyperspace subserves. Reply, partly explicit but chiefly implicit, is not, I trust, entirely wanting in what has been already said. Motivity, at all events, and *raison d'être* are not far to seek. On the one hand, the great generalization has made it possible to enrich, quicken and beautify analysis with the terse, sensuous, artistic, stimulating language of geometry. On the other hand, the hyperspaces are in themselves immeasurably interesting and inexhaustibly rich fields of research. Not only does the geometrician find light in them for the illumination of many otherwise dark and undiscovered properties of the ordinary spaces of intuition, but he also discovers there wondrous structures quite unknown to ordinary space. These he examines. He handles them with the delicate instruments of his analysis. He beholds them with the eye of the understanding and delights in the presence of their supersensuous beauty.

Creation of hyperspaces is one of the ways by which the rational spirit secures release from limitation. In them it lives ever joyously, sustained by an unfailing sense of infinite freedom.[4]

---

[4] Readers wishing to pursue the matter further will find a fresh and fuller discussion in the chapter on Hyperspace in Keyser's *Mathematical Philosophy,* Dutton and Company. This work is designed for educated laymen.

THE UNIVERSE AND BEYOND: THE EXISTENCE OF THE HYPERCOSMIC [1]

Ni la contradiction n'est marque de fausseté, ni l'incontradiction n'est marque de vérité. — PASCAL

THE inductive proof of the doctrine of evolution seems destined to be ultimately judged as the great contribution of Natural Science to modern thought. Among the presuppositions of that doctrine, among the axioms, as one may call them, of science, are found the following: —

(1) The assumption of the universal and eternal reign of *law:* the assumption that the universe, the theatre of evolution, the field of natural science, is and eternally has been a genuine Cosmos, an incarnate rational logos, an embodiment of reason, an organic affair of order, a closed domain of invariant uniformities, in which waywardness and chance have had nor part nor lot: an infinitely intricate garment, ever changing, yet always essentially the same, woven, warp and weft alike, of mathetic relationships.

(2) The assumption, not merely that the universe is cosmic through and through, but that it is the *all* conjunctively — the all, that is, in the sense of naught excluded; the assumption, in other words, that it is not merely *a* but *the* cosmos, the *sole* system of law and order and harmony, the complete and perfect embodiment of the *whole* of truth.

Such, I take it, are among the principles, the articles of faith, more or less consciously held by the great majority of the men of science and their adherents.

[1] Appeared in *The Hibbert Journal*, January, 1905.

As for myself, I am unable to hold these tenets either as self-evident truths, or as established facts, or as propositions the proof of which may be confidently awaited. Truth, for example, especially when contemplated in its relations to curiosity — at once *the* psychic product and psychic agency of evolution — less seems a completed thing coeval with the world than a thing derived and still becoming. Again, while the assumption of the cosmic character of our universe is of the greatest value as a working hypothesis, I am unable to find in the method of natural science or in that of mathematics any ground, even the slightest, for expecting conclusive proof of its validity. In striking contrast, on the other hand, with this negative thesis, there is found in the realm of pure thought, in the domain of mathematics, very convincing evidence, not to say indubitable proof of the proposition, that no single cosmos, whether our universe be such or not, can enclose every rationally constructible system of truth, but that any universe is a component of an extra-universal, that above every nature is a super-natural, beyond every cosmos a hypercosmic.

These are among the theses presented in the following pages, not in a controversial spirit, let me add, nor accompanied by the minuter arguments upon which they ultimately rest.

We all must allow that truth is. To deny it denies the denial. Such scepticism is cut away by the sweeping blade of its own unsparing doubt. But *what* it is — that is another matter. The assumption that truth is an agreement or correspondence between concepts and things, between thought and object, is of very great value in practical affairs; it very well serves, too, the *immediate* purposes of natural science, especially in its cruder stage,

before it has learned by critical reflection on its own processes and foundations to suspect its limitations, and
while, like the proverbial " chesty " youth who disdains
the meagre wisdom of his father, it is apt to proclaim,
innocently enough if somewhat boastfully, a lofty con-
tempt for all philosophy and metaphysics. Although the
assumption has the undoubted merit of being thus useful
in high degree, it is, when regarded as a definitive formulation of what we mean by truth, hardly to be accepted.
For, not only does it imply — what may indeed be quite
correct, but is far from being demonstrated, and far from
being universally allowed — namely, that " thing " is one
and " concept " another, that " object " and " thought "
are twain, but even if we grant such ultimate implied
duality, it remains to ask what that " agreement " is,
or " correspondence," that mediates the hemispheres and
gives the whole its truth. The assumption is slightly too
naïve and unsophisticated, a little too redolent of an un-
tamed soil and primitive stage of cultivation. Much
profounder is that insight of Hegel's, that truth is the
harmony which prevails among the objects of thought.
If, with that philosopher, we identify object and thought,
we have at once the pleasing utterance that truth is the
harmony of ideas. But here, again, easy reflection
quickly finds no lack of difficulties. For what should we
say an idea is? And *is* there really nothing else, except, of
course, their harmony? And what is that? And is there
no such thing as contradiction and discord? Is that, too,
a kind of truth, a kind of harmonious jangling, a melody
of dissonance? The fact seems to be that truth is so
subtle, diverse, and manifold, so complex of structure and
rich in aspect, as to defy all attempt at final definition.
Nay, more, the difficulty lies yet deeper, and is in fact
irresoluble. Being a necessary *condition* thereto, truth

can not be an *object* of definition. To suppose it defined
involves a contradiction, for the definition, being something new, is something *besides* the truth defined, but it
must itself be true, and, if it be, in that has failed — the
enclosing definition is not itself enclosed, and straightway asks a vaster line to take it in, and so *ad infinitum*.
To define truth would be to construct a formula that
should include the structure, to conceive a water-com-
passed ocean, bounded in but shutting nothing out, a
self-immersing sea, without bottom or surface or shore.

Happily, to be indefinable is not to be unknowable
and not to be unknown. And we are absolutely certain
that truth, whatever it may be, is somehow the *complement of curiosity,* is the proper stuff, if I may so express
it, to answer questions with. Now a question, once one
comes to think of it, is a rather odd phenomenon. Half
the secret of philosophy, said Leibniz, is to treat the
familiar as unfamiliar. So treated, curiosity itself is a
most curious thing. How blind our familiar assumptions
make us! Among the animals, man, at least, has long
been wont to regard himself as a being quite apart from
and not as part of the cosmos round about him. From
this he has detached himself in thought, he has estranged
and objectified the world, and lost the sense that he is of
it. And this age-long habit and point of view, which has
fashioned his life and controlled his thought, lending its
characteristic mark and colour to his whole philosophy
and art and learning, is still maintained, partly because
of its convenience no doubt, and partly by force of inertia
and sheer conservatism, in the very teeth of the strongest
probabilities of biologic science. Probably no other
single hypothesis has less to recommend it, and yet no
other so completely dominates the human mind. Suppose
we deny the assumption, as we seem indeed compelled

to do, in the name of science, and readjoin ourselves in thought, as we have ever been joined in fact, to this universe in which we live and have our being; the other half of the secret of philosophy will be revealed, or illustrated at all events, in the strangeness of aspect presented by things before familiar. Note the radical character of the transformation to be effected. The world shall no longer be beheld as an alien thing, beheld by eyes that are not its own. Conception of the whole and by the whole shall embrace *us* as *part,* really, literally, consciously, as the latest term, it may be, of an advancing sequence of developments, as occupying the highest rank perhaps in the ever-ascending hierarchy of being, but, at all events, as emerged and still emerging *natura naturata* from some propensive source within. I grant that the change in point of view is hard to make — old habits, like walls of rock, tending to confine the tides of consciousness within their accustomed channels — but it can be made and, by assiduous effort, in the course of time, maintained. Suppose it done. By that reunion, the whole regains, while the part retains, the consciousness the latter purloined. I cannot pause to note even the most striking consequences of such a change in point of view. Time would fail me to follow far the opening lines of speculation that issue thence and invite pursuit. But I cannot refrain from pointing out how exceedingly curious a thing curiosity itself becomes when beheld and contemplated from the mentioned point of view. For it is now the whole that meditates, the universe that contemplates — a once *mindless* universe according to its present understanding of the term, not then knowing that it was, unwittingly unwitting throughout a beginningless eternal past what it had been or was or was to be; lawless, too, perhaps, could the stream of events be reascended, though blindly and slowly *becoming* lawful through habit-taking tendence: a self-transforming insensate mass composed of parts without likeness or distinction, continually undergoing change without a purpose, devoid of passion, and neither ignorant nor having knowledge. At length a wondrous crisis came, an event momentous — when or how is yet unknown, perhaps through fortuitous concourse of partless, lawless, wayward elements. At all events, the unintending tissues formed a nerve, the universe awoke alive with wonder, mind was born with curiosity and began to look about and make report of part to part and thence to whole, the age of interrogation was at hand, and what had been an eternal infinity of mindless being began to question, and know itself, and have a sense of ignorance. In the whole universe of events, none is more wonderful than the birth of wonder, none more curious than the nascence of curiosity itself, nothing to compare with the dawning of consciousness in the ancient dark and the gradual extension of psychic life and illumination throughout a cosmos that before had only *been*. An eternity of blindly acting, transforming, unconscious existence, assuming at length, through the birth of sense and intellect, without loss or break of continuity, the abiding form of fleeting time. Another eternity remains to follow, and one cannot but wonder whether there shall issue forth in future from the marvel-weaving loom another event, or form or mode of being, that shall be to the modern universe that both is and knows, as the birth of soul and curiosity to the ancient universe that was but did not know. A speculation by no means idle, but let it pass.

I wish to point out next, briefly, that curiosity is not only a principle that leads to knowing, but a principle and process of growing. By it the universe comes not merely to understand itself, but actually to get bigger thereby. For if there be an invariant amount of matter, there is also mind increasing; if there be objects that total a constant sum, there are also ideas that multiply. A new query and a new answer are new elements in the world, by which the latter is added unto and enriched. Curiosity is the aspect of the universe seeking to realise itself, and the fruit of such activity is new reality, stimulating to new research. Imagine a body with an inner core of outward-striving impulses producing buds at every radial terminus. Such is knowledge — a kind of proliferating sphere, expanding along divergent lines by the outward-seeking of an inner life of wonder. Wherefore, it appears again that truth, the complement of curiosity, itself grows with the latter's growth, and, being never a finished thing, but one that both is and is becoming, is not to be compassed by definition nor fully solved in knowledge.

In respect to truth, then, the upshot is: we are certain that it is; not, however, as a closed or completed scheme of relationships, but as a kind of reality characterised by the phenomena of growth and of becoming; it does not admit of ultimate definition; we know, however, in a super-verbal sense, through myriad manifestations of it to a faculty in us of *feeling* for it, what it is; we recognise it as the motive power, the *elixir vitæ*, the sustaining spring of wonder; it discovers itself as the wherewithal for the proper fulfilment of the implicit predictions and intimations of curiosity; as *the* thing presaged in a spiritual craving, confidently, persistently proclaiming its needs by an infinitude of questionings.

And now as to the remainder of my subject, the tale is quite too long to be told in full. But room must be found for a partial account, for important fragments at all events.

What, then, shall we say mathematics is? A question much discussed by philosophers and mathematicians in the course of more than two thousand years, and especially with deepened interest and insight in our own time. Many an answer has been given to it, but none has approved itself as final. Naturally enough, conception of the science has had to grow with the science itself. For it must not be imagined that mathematics, because it is so old, is dead. Old it is indeed, classic already in Euclid's day, being surpassed in point of antiquity by only one of the arts and by none of the sciences; but it is also living and new, flourishing to-day as never before, advancing in a thousand directions by leaps and bounds. It is not merely as a giant tree throwing out and aloft myriad branching arms in the upper regions of clearer light, and plunging deep and deeper roots in the darker soil beneath. It is rather an immense forest of such oaks, which, however, literally grow into each other, so that, by the junction and intercresence of root with root and limb with limb, the manifold wood becomes a single living organic whole. A vast complex of interlacing theories — *that* the science now is actually, but it is far more wondrous still potentially, its component theories continuing more and more to grow and multiply beyond all imagination, and beyond the power of any single genius, however gifted. What is this thing so marvellously vital? What does it undertake? What is its motive? How is it related to other modes and interests of the human spirit?

One of the oldest and at the same time the most familiar of the definitions conceived mathematics to be the science of magnitude, where magnitude, including multitude as a special kind, was whatever was capable of increase and decrease and measurement. This last — capability of measurement — was the essential thing. That was a most natural definition of the science, for magnitude is a singularly fundamental notion, not only inviting but demanding consideratior at every stage and turn of life. The necessity of finding out how many and how much was the mother of counting and measurement, and mathematics, first from necessity and then from joy, so busied itself with these things that they came to seem its whole employment. But now the notion of measurement as the repeated application of a constant unit has been so refined and generalised, on the one hand through the creation of imaginary and irrational numbers, and on the other by use of a scale, as in non-Euclidian geometry, where the unit suffers a lawful change from step to step of its application, that to retain the old words and call mathematics the science of measurement seems quite inept as no longer telling what the spirit of mathesis is really bent upon. Moreover, the most striking measurements, as of the volume of a planet, the swiftness of thought, the valency of atoms, the velocity of light, the distance of star from star, are not achieved by direct repeated application of a unit. They are all accomplished by *indirection*. And it was perception of this fact which led to the famous definition by the philosopher and mathematician, Auguste Comte, that mathematics is the science of *indirect* measurement. Doubtless we have here a finer insight and a larger view, but the thought is yet too narrow, nor is it deep enough. For it is obvious that there is much mathematical activity which is not at all concerned with measurement, either direct or indirect. In projective geometry, for example, it was observed

that *metric* considerations were by no means chief. As
a simplest illustration, the fact that two points determine a line, or the fact that a plane cuts a sphere in a
circle, is not a *metric* fact, being concerned with neither
size nor magnitude. Here it was position rather than
size that seemed to some to be the central idea, and so it
was proposed to call mathematics the science of magnitude, or measurement, and *position*.

Even as thus expanded, the definition yet excludes
many a mathematical realm of vast, nay, infinite extent.
Consider, for example, that immense class of things familiarly known as *operations*. These are limitless, alike
in number and in kind. Now it so happens that there
are systems of operations such that any *two* operations
of a given system which follow one another produce the
same effect as some other *single* operation of the system.
For an illustration, think of all possible straight motions
in space. The operation of going from A to B followed
by the operation of going from B to C is equivalent to
the single operation of going from A to C. Thus, the
system of such straight operations is a *closed* system.
Combination of any two of them yields another operation,
not without, but *within* the system. Now the theory of
such closed systems — called groups of operations — is a
mathematical theory, already of colossal proportions, and
still growing with astonishing rapidity. But, and this is
the point, an abstract operation, though a very real thing,
is neither a position nor a magnitude.

This way of trying to come at an adequate conception
of mathematics, *viz.*, by naming its different domains,
or varieties of content, is not likely to prove successful.
For it demands an exhaustive enumeration not only of
the fields now occupied by the science, but also of the
realms destined to be conquered by it in the future, and

such an achievement would require a prevision that none perhaps could claim.

Fortunately there are other paths of approach that seem more promising. Everyone has observed that mathematics, whatever it may be, possesses a certain mark, namely, a degree of certainty not found elsewhere. So it is, proverbially, the exact science *par excellence*. Exact, you say, but in what sense? To this an excellent answer is contained in a definition given by an American mathematician, Professor Benjamin Peirce: *Mathematics is the science which draws necessary conclusions,* a formulation something more than finely paraphrased by one [2] of my own teachers thus: *Mathematics is the universal art apodictic.* These statements, though neither of them may be entirely satisfactory, are both of them telling approximations. Observe that they place the emphasis on the quality of being *correct*. Nothing is said about the conclusions being true. That is another matter, to which I will return presently. But why are the conclusions of mathematics correct? Is it that the mathematician has an essentially different reasoning faculty from other folks? By no means. What, then, is the secret? Reflect that conclusion implies premises, and premises imply terms, and terms stand for ideas or concepts, and that *these,* namely, concepts, are the *ultimate* material with which the spiritual architect, which we call the Reason, designs and builds. Here, then, we may expect to find light. The apodictic quality of mathematical thought, the certainty and correctness of its conclusions, are due, not to a special mode of ratiocination, but to the character of the concepts with which it deals. What is that distinctive characteristic? I answer: *precision, sharpness, completeness, of definition.*

[2] Professor W. B. Smith.

But how comes your mathematician by such completeness? There is no mysterious trick involved; some ideas
admit of such precision, others do not; and the mathematician is one who deals with those that do. Law, says
Blackstone, is a rule of action prescribed by the supreme
power of a state commanding what is right and prohibit-
ing what is wrong. But what are a state and supreme
power and right and wrong? If all such terms admitted
of complete determination, then the science of law would
be a branch of pure, and its practice a branch of applied,
mathematics. But does not the lawyer sometimes arrive
at correct conclusions? Undoubtedly he does sometimes,
and, what may seem yet more astonishing, so does your
historian and even your sociologist, and that without the
help of accident. When this happens, however, when
these students arrive, I do not say at truth, for that may
be by lucky accident or happy chance or a kind of intuition, but when they arrive at *conclusions* that are *correct,*
then that is because they have been for the moment in
all literalness acting the part of mathematician. I do not
say that for the aggrandisement of mathematics. Rather
is it for credit to *all* thinkers that none can show you any
considerable garment of thought in which you may not
find here and there, rarely enough sometimes, a golden
fibre woven in some, it may be, exceptional moment, of
precise conception and rigorous reasoning. To think
right — that is no characteristic striving of a class of
men. It is a common aspiration. Only, the stuff of
thought is mostly intractable, formless, like some milky
way waiting to be analysed into distinct star-forms of
definite ideas. All thought aspires towards the character
and condition of mathematics.

The reality of this aspiration and the distinction it
implies admit of many illustrations, of which here a single

one must suffice. There is no more common or more important notion than that of *function*, the term being applied to either of two variable things such that to any value or state of either there correspond one or more values or states of the other. Of such function pairs, examples abound on every hand, as the radius and the area of a circle, the space traversed and the rate of going, progress of knowledge and enthusiasm of study, elasticity of medium and velocity of sound or other undulation, the amount of hydrogen chloride formed and the time occupied, the prosperity of a given community and the intelligence of its patriotism. Indeed, it may very well be that there is nothing which is not in some sense a function of every other. Be that as it may, one thing is very certain, namely, a very great part and probably all of our thinking is concerned with functional relationships, deals, that is, with pairs of systems of corresponding values or states or changes. Behold, for example, how the parallelistic psychology searches for correlations between psychical and physical phenomena. Witness, too, the sociologist trying to determine the correspondence between the peacefulness and the homogeneity of a population, or, again, between manifestations of piety or the spread of populism and the condition of the crops. It is then here, in the wondrous domain of correspondence, the answering of value to value, of change to change, of condition to condition, of state to state, that the knowing activity finds its field.

What is it precisely that we seek in a correlation? The answer is: *when one or more facts are given, to pass, with absolute certainty, to the correlative fact or facts.* To do this obviously requires formulæ or equations which precisely define the manner of correlation, or the law of interdependence. Where do such formulæ come from?

I answer that, strictly speaking, they are never *found*, they are always *assumed*. Now, nothing is easier than to write down a perfectly definite formula that does not tell, for example, how cheerfulness depends on climate, or how pressure affects the volume of a gas. Nay, a given formula may be perfectly intelligible in itself, it may state, that is, a perfectly intelligible law of correspondence, which, nevertheless, may have no validity at all in the physical universe and none elsewhere than in the formula itself. What, then, guides in the choice of formulæ? That depends upon your kind of curiosity, and curiosity is not a matter of choice.

Just here we are in a position where we have only to look steadily a little in order to see the sharp distinction between mathematics and natural science. These are discriminated according to the kind of curiosity whence they spring. The mathematician is curious about definite abstract correspondences, about perfectly-defined functional relationships *in themselves*. These are more numerous than the sands of the seashore, they are as multitudinous as the points of space. It is this assemblage of pure, precisely-defined relationships which constitute the mathematician's universe, an indefinitely infinite universe, worlds of worlds of wonders, inconceivably richer than the outer world of sense. This latter is indeed immense and marvellous, with its rolling seas and stellar fields and undulating ether, but, compared with the hyperspaces explored by the genius of the geometrician, the whole vast extent of the sensuous universe is a merest point of light in a blazing sky.

Now this mere speck of a physical universe, in which the chemist and the physicist, the biologist and the sociologist, and the rest of nature devotees, find their great fields, may be, as it seems to be, an organic thing, connected into an ordered whole by a tissue of definable functional relationships, and it may not. The nature devotee *assumes* that it is and *tries* to find the relationships. The mathematician does *not* make that assumption and does *not* seek for relationships in the *outer* world. Is the assumption correct? As man, the mathematician does not know, although he greatly cares. As mathematician, man neither knows nor cares. The mathematician does know, however, that, *if* the assumption *be* correct, every definite relationship that is valid in nature, every type of order and mode of correlation obtaining there, is, in itself, a thing for his thought, an essential element in his domain of study. He knows, too, that, if the assumption be *not* correct, his domain remains the same absolutely. The two realms, of mathematics, of nature science, are fundamentally distinct and disparate forever. To think the thinkable — that is the mathematician's aim. To assume that nature is thinkable, an incarnate rational logos, and to seek the thought supposed incarnate there — these are at once the principle and the hope of the nature student. Science, said Riemann,[3] is the *attempt* to comprehend nature by means of *concepts*. Suppose the nature student is right, suppose the physical universe really is an enfleshed logos of reason, does that imply that *all* the thinkable is thus incorporated? It does not. A single ordered universe, one that through and through is self-compatible, cannot be the whole of reason materialised and objectified. *There is many a rational logos,* and the mathematician has high delight in the contemplation of *in*consistent *systems* of *consistent relationships*. There are, for ex-

---

[3] *Cf.* Riemann: "Fragmente Philosophischen Inhalts," in *Gesammelte Werke.* These fragments, which are published in English by the Open Court Pub. Co., Chicago, are exceedingly suggestive.

ample, a Euclidean geometry and more than one species of non-Euclidean. As theories of a *given* space, these are not compatible. If our universe be, as Plato thought, and nature science takes for granted, a space-conditioned, geometrised affair, one of these geometries may be, none of them may be, not all of them can be, valid in it.. But in the vaster world of thought, all of them are valid, there they co-exist, and interlace among themselves and others, as differing component strains of a higher, strictly supernatural, hypercosmic, harmony.

It is, then, in the inner world of pure thought, where all *entia* dwell, where is every type of order and manner of correlation and variety of relationship, it is in this infinite ensemble of eternal verities whence, if there be one cosmos or many of them, each derives its character and mode of being, — it is there that the spirit of mathesis has its home and its life.

Is it a restricted home, a narrow life, static and cold and grey with logic, without artistic interest, devoid of emotion and mood and sentiment? That world, it is true, is not a world of *solar* light, not clad in the colours that liven and glorify the things of sense, but it is an illuminated world, and over it all and everywhere throughout are hues and tints transcending *sense,* painted there by·radiant pencils of *psychic* light, the light in which it lies. It is a silent world, and, nevertheless, in respect to the highest principle of art — the interpenetration of content and form, the perfect fusion of mode and meaning — it even surpasses music. In a sense, it is a static world, but so, too, are the worlds of the sculptor and the architect. The figures, however, which reason constructs and the mathematical vision beholds, transcend the temple and the statue, alike in simplicity and in intricacy, in delicacy and in grace, in symmetry and in poise. Not

only are this home and this life thus rich in æsthetic
interests, really controlled and sustained by motives of a
sublimed and supersensuous art, but the religious aspiration, too, finds there, especially in the beautiful doctrine
of invariants, the most perfect symbols of what it seeks —
the changeless in the midst of change, abiding things in
a world of flux, configurations that remain the same despite the swirl and stress of countless hosts of curious
transformations. The domain of mathematics is the sole
domain of certainty. There and there alone prevail the
standards by which every hypothesis respecting the external universe and all observation and all experiment
must be finally judged. It is the realm to which all
speculation and all thought must repair for chastening
and sanatation — the court of last resort, I say it rever-
ently, for all intellection whatsoever, whether of demon
or man or deity. It is there that mind as mind attains
its highest estate, and the condition of knowledge there
is the ultimate object, the tantalising goal of the aspiration, the *Anders-Streben,* of all other knowledge of every
kind.

THE AXIOM OF INFINITY: A NEW
PRESUPPOSITION OF THOUGHT [1]

It so happened that when the first number of *The Hibbert Journal* appeared, containing an article by Professor Royce on the Concept of the Infinite, I had been myself for some time meditating on the logical bearings and philosophical import of that concept, and was actually then engaged in marking out the course which it seemed to me a first discussion of the matter might best follow. The order and scope of his treatment were so like those I had myself decided upon that I should naturally have felt a pardonable pride in the coincidence, had not this feeling been at the same time quite lost in a stronger one, namely, that of the evident superiority of his manner to any which I could have hoped to attain. Indeed, so patient is his exposition of elements, so rich is it in suggestiveness, so intimately and instructively, according to his wont, has he connected the most abstruse and recondite of doctrines with the most obvious and seemingly trivial of things, and so luminous and stimulating is it all, that one must admire the ingenuity it betrays, and cannot but wonder whether after all there really are in science or philosophy any notions too remote and obscure to be rendered intelligible even to common sense, if only a sufficiently cunning pen be engaged in the service.

While his paper is thus replete with inspiring intimations of the "glorious depths" and near-lying interests of

[1] Appeared in *The Hibbert Journal*, April, 1904.

the doctrine treated, and is, in point of clearness and vivid portrayal of its central thought, a model beyond the art of most, it is not, I believe, equally happy when judged on the severer ground of its critico-logical estimates. Even on this ground, I do not hesitate, after close examination, to adjudge it the merit of *general* soundness. That, however, it is thoroughly sound, completely mailed against every possible assault of criticism, is a proposition I am by no means prepared to maintain. Quite the contrary, in fact. Nor can the defects be counted as trivial. One of them especially, which it has in common with other both earlier and later discussions of the subject, notably that by Dedekind himself and, more recently, that by Mr. Bertrand Russell in his imposing treatise on *The Principles of Mathematics,* is of the most radical nature, concerning as it does no less a question than, I do not say merely that of the validity, but that of the possibility, of existence-proofs of the infinite.

And here I may as well state at once, lest there should be some misapprehension in respect to purpose, that the present writing is not primarily designed to be a review of Professor Royce's or of other recent discussions of the infinite. Reviewed to some extent they will be, but only incidentally, and mainly because they have declared themselves, erroneously as I think, upon that most fundamental of questions, namely, *whether it is possible, by aid of the modern concept, to demonstrate the existence, of the infinite.* Argument would seem superfluous to show the immeasurable import of this problem, whether it be viewed solely in its immediate logical bearings, or also mediately, through the latter, in its bearings upon philosophy, upon theology, and, only more remotely, upon religion itself. It is chief among the aims of this essay, to open that problem anew, to appeal from the prevailing doctrine concerning it, in the hope of securing, if possible, a readjudication of the matter which shall be final.

This subject of the infinite, how it baffles approach! How immediate and how remote it seems, how it abides and yet eludes the grasp, how familiar it appears, mingling with the elemental simplicities of the heart, continuously weaving itself into the intimate texture of common life, and yet how austere and immense and majestic, outreaching the sublimest flights of the imagination, transcending the stellar depths, immeasurable by the beginningless, endless chain of the ages! Comprehend the infinite! No wonder we hear that none but the infinite itself is adequate to that. *Du gleichst dem Geist, den du begreifst.* Be it so. Perhaps, then, we are infinite. If not,

> "'Wie' fass' ich dich, unendliche Natur?'"

Or is it finally a mere illusion? And is there after all no infinite reality to *be* seized upon? Again, if not, what signifies the finite? Is that to be for ever without definition, except as reciprocal of that which fails to be? Is the All really enclosed in some vast ellipsoid, without a beyond, incircumscriptible, devoid alike of tangent plane and outer point? Are we eternally condemned to seek therein for the meaning and end of processes that refuse to terminate? And is, then, this region, too, but a locus of deceptions, " of false alluring jugglery " ? Is analysis but the victim of hallucination when it thinks to detect the existence of realms that underlie and overarch and compass about the domain of the countable and measurable? And does the spirit, in its deeper musings, in its pensive moods, only *seem* to feel the tremulous touch of transfinite waves, of vitalising undulations from beyond the farthest shore of the sea of sense?

One fact at once is clear, namely, that, whatever ultimate justification the hypothesis may find, thought has never escaped the necessity of *supposing* the universe of things to be intrinsically somehow cleft asunder into the two Grand Divisions, or figured, if you will, under the two fundamental complementary all-inclusive Forms, which, from motives more or less distinctly felt and also just, as we shall see, though not quite justified, have been, from time immemorial, designated as the Finite and the Infinite. And these great *terms* or their verbal equivalents — for concepts in any strict sense they have not been — though always vague and shifting, for ever promising but never quite delivering the key to their identities into the hand of Definition, have, nevertheless, in every principal scene, together played the gravest *rôle* in the still unfolding drama of speculation. Or, to change the figure, they have been as Foci, one of them seemingly near, the other apparently remote, neither of them quite itself determinate, but the two conjointly serving always to determine the ever-varying eccentricity of the orbit of thought; and doubtless the vaster lines that serve to bind the differing epochs of speculation into a single continuous system can best be traced by reference to these august terms as co-ordinate poles of interest.

As a simple historical fact, then, philosophy has indeed, with but negligible exception, throughout *assumed* the existence of both the finite and the infinite. That is one thing. Another fact of distinct and equal weight, no matter whether or how we may account for it, is that man, in accord with the deeper meaning of the Protagorean maxim, has always felt himself to have within, or to be somehow, the potential measure of all that is. Is it insignificant that this faith — for that is what it seems to be — as if an indestructible character of the race, as if an invariant defining property of the germ plasm itself whence man springs and derives his continuity, should have survived every vicissitude of human fortune? that it should have been indeed, if not the substance, at least the promise, of things hoped for, the evidence, too, of things not seen, marking and sustaining metaphysical research from the earliest times? And, what is more, the spirit of such research, curiosity I mean, fit companion and counterpart of that abiding faith, unlike " experience and observation," has known no bounds, but, on the contrary, finding within itself no fatal principle of limitation, it has ever disdained the scale of finite things as competent to take its measure, and boldly asserted claim to the entire realm of being.

These questions, however, have been something more than fascinating. Perhaps their rise, but not their manifold development, much less their profound significance for life and thought, is to be adequately explained on the hypothesis of insatiate curiosity alone. It must be granted that their presence, especially in the arena of dialectic, *has* been often due simply to their intrinsic magical charm for " summit-intellects." And doubtless the play-instinct, deep-dwelling in the constitution of the mind, has often made them serve the higher faculties merely as intricate puzzles, to beguile the time withal. But, in general, the questions have worn a sterner aspect. Philosophy has been not merely allured, it has been constrained, to their consideration; constrained not only because of their inherence in problems of the conscience, especially in that most radical problem of finding the simplest system of postulates that shall be at once both necessary and sufficient to explain the moral feeling; but constrained still more powerfully by the insistent de-

mands that issue from the religious consciousness. But
this is yet not all. For man cannot live by these august
interests alone. And it is profoundly significant, both as
witnessing to the final inter-blending, the fundamental
unity, of all the concerns of the human spirit, and as revealing the ultimate depth and dignity of all its interests,
that questions about the infinite quite similar to those
that claim so illustrious parentage in Ethics and Philosophy, admit elsewhere of humbler derivation, and readily
own to the lowliest of origins. Man, indeed, merely to
live, has had to measure and to count, and this homely
necessity, fruitful mother of mystery and doubt, *independently* set the problems of the indefinitely small and
the indefinitely great; and so it was that needs quite as
immediate and austere as those of Morals and Religion
— I mean the exigencies of Science, and especially of
Mathematics — demanded on their own ground, in the
very beginnings of exact knowledge, that the understanding transcend every possible sequence of observations,
pass the uttermost limit of " experience," which, refine
and enlarge it as you may, remains but finite, and literally lay hold on infinity itself.

To this ancient irrevocable demand, thus urged upon
the reason from every cardinal point of human interest,
genius has responded as to a challenge from the gods,
and I submit that the response, the endeavour of the
reason actually to subjugate extra-finite being and compel surrender of its secrets by the organon of thought,
constitutes the most sublime and strenuous and inspiring
enterprise of the human intellect in every age.

What of it? Long centuries of gigantic striving, age
on age of philosophic toil, immeasurable devotion of
time and energy and genius to a single end, the intellectual conquest of transfinite being — what has it all

availed? What triumphs have been won? I speak, narrowly, of the conquest, and demand to know, not whether it has been accomplished — for that were a foolish query — but whether, strictly speaking, it has been begun. Let not the import of the question be mistaken. No answer is sought in terms of such moral or " spiritual " gains as may be incident even to efforts that miss their aim. Everyone knows that seeking has compensations of its own, which indeed are ofttimes better than any which finding itself can give. And it seems sometimes as if the higher life were chiefly sustained by unsought gains incident to the unselfish pursuit of the unattainable. The circle has not been squared, nor the quintic equation solved,[2] nor perpetual motion invented; neither indeed can be; yet it would show but meagre understanding of the ways of truth to men, did one suppose all the labour devoted to such problems to have been without reward. So, conceivably, it might be with this problem of the infinite. It may be granted that, even supposing no solution to be attainable, the ceaseless search for one, the unwearied high endeavor of the reason through the ages, presents a spectacle ennobling to behold, and of which mankind, it may be, could ill afford to be deprived. It may be granted that incidentally many insights have been won which, though not solutions, have nevertheless permanently enriched the literature of the world and are destined to improve its life. It may be granted that in every time some doctrine of infinity, some philosophy of it, has been at least effective, has helped, that is, for better or worse, to fashion the forms of human institutions and to determine the course of history. Concerning none of these things is there here any question. As to what the

[2] That is, by means of radicals.

question precisely is, there need not be the slightest misapprehension.   The fact is that for thousands of years
philosophy has recognised the presence of a certain definite Problem, namely, that of *extending the dominion
of logic, the reign of exact thought, out beyond the utmost
reach of finite things into and over the realm of infinite
being,* and this problem, by far the greatest and most
impressive of her strictly intellectual concernments,
philosophy has, for thousands of years, arduously striven
to solve.   And now I ask — not, has it been worth while?
for that is conceded, but — has she advanced the *solution* in any measure, and, if so, in what respect, and to
what extent?

We are here upon the grounds of the *rational* logos.
The whole force and charge of the question is directed
to matter of concept and inference.   Fortunately, the answer is to be as unmistakable as the question.   It must
be recognised, of course, that the " problem," as stated,
is exceedingly, almost frightfully, generic, comprising
a host of interdependent problems.   One of these, however, is pre-eminent: without its solution *none* other
can be solved; with its solution, *any* other *may* be eventually.   That problem is the problem of conception, of
definition in the unmitigated rigour of its severest meaning; it is the problem of discovering a certain principle,
of finding, without the slightest possibility of doubt or
indetermination, the intrinsic line of cleavage that parts
the universe of being into its two grandest divisions, and
so of telling finally and once for all precisely what, for
thought, the infinite is and what, for thought, the finite is.

And now, thanks to the subtle genius of the modern
Teutonic mind, this ancient problem, having baffled the
thought of all the centuries, has been at last completely
solved, and therein our original question finds its answer:

The conquest has been begun.   Bernhard Riemann, profound mathematician and — important fact, of which, strangely enough, too many philosophers seem invincibly unaware — profound metaphysician too, having pointed out, in his famous *Habilitationschrift,*[3] the epoch-making distinction between mere boundlessness and infinitude of manifolds similar to that of space, the greater glory was reserved for three contemporary compatriots of his — Bernhard Bolzano,[4] Richard Dedekind,[5] and Georg Cantor,[6] the first an acute and learned philosopher and theologian, with deep mathematical insight, the other two brilliant mathematicians, with a strong bent for metaphysics — to win independently and about the same time the long-coveted insight into the intrinsic nature of infinity.   And thus it is a distinction of our own time that within the memory of living men the *defining mark* of the infinite first failed to elude the grasp, and that august term, after the most marvellous career of any in the history of speculation, has been finally made to assume the prosaic form of an exact and completely determined concept, and so at length to become available for the purposes of rigorously logical discourse.

Pray, then, what is this concept?   Of various equivalent forms of statement, I choose the following: *An assemblage (ensemble, collection, group, manifold) of elements (things, no matter what) is infinite or finite according as it has or has not a* PART *to which the whole is just* EQUIVALENT *in the sense that between the elements composing that part and those composing the whole*

---

[3] " Ueber die Hypothesen, welche die Geometrie zu Grunde liegen," *Ges. Werke.* Also in English by W. K. Clifford.

[4] " Paradoxien des Unendlichen."

[5] " Was sind und was sollen die Zahlen."

[6] Memoirs in *Acta Mathematica,* vol. ii., and elsewhere.

there subsists a unique and reciprocal (one-to-one) correspondence.

If we may trust intuition in questions about reality, assemblages,[7] infinite as defined, actually abound on every hand. I need not pause to indicate examples. Those pointed out in Professor Royce's mentioned paper may suffice; they will, at all events, furnish the reader with the " clew, which, once familiar to his hand, will lengthen as he goes, and never break." The concept itself I regard as a great achievement, one of the very greatest in the history of thought. Not only does it mark the successful eventuation of a long and toilsome search; it furnishes criticism with a new standard of judgment, it at once creates, and gives the means of meeting, the necessity for a re-examination and a juster evaluation of historic doctrines of infinity; and it is greater still, I believe, as a destined instrument of exploration in that realm which it has opened to the understanding and whose boundary it defines.

Is that judgment not extravagant? For the concept seems so simple, is so apparently independent of difficult presuppositions, that one cannot but wonder why it was not formed long ago. Had the concept in question been early formed, the history and present status of philosophy and theology, and of science too, had doubtless been different. But it was not then conceived. Now that we have it, is it too unbewildering to be impressive? Shall we esteem it lightly just because we can comprehend it, because it does not mystify? Simple it is indeed, almost as simple as the Newtonian law of

---

[7] The very simplest possible example of such a manifold is that of the count-numbers. The whole collection can be paired in one-to-one fashion with, for example, half the collection, thus: 1, 2; 2, 4; 3, 6; . . . ; the totality of even and odd being just equivalent to the even.

gravitation, nearly as easy to understand as the geometric interpretation of imaginary quantities, hardly
more difficult to grasp than the notion of the conservation of energy, the Mendelian principle of inheritance,
or than a score of other central concepts of science.
But shallow indeed and foolish is that criticism which
values ideas according to their complexity, and con-
founds the simple with the trivial.

As an immense city or a vast complex of mountain
masses, seen too near, is obscured as a whole by the
prominence of its parts, so the larger truth about any
great subject is disclosed only as one beholds it at a
certain remove which permits the assembling of principal
features in a single view, and a proportionate mingling
of reflected light from its grander aspects. Accordingly
it has seemed desirable, in the foregoing preliminary
survey, to hold somewhat aloof, to conduct the movement, in the main along the path of perspective centres,
in order to allow the vision at every point the amplest
range. It is now proposed to draw a little closer to the
subject and to examine some of its phases more minutely.
In respect to the modern concept of infinity, we desire
to know more fully what it really signifies, we wish
to be informed how it orients itself among cardinal
principles and established modes of thought. But recently born to consciousness, it has already been advanced to conspicuous and commanding station among
fundamental notions, and we are concerned to know
what, if any, transformations of existing doctrine,
what readjustments of attitude towards the universe
without us or within, what changes in our thought on
ultimate problems of knowledge and reality, it seems to
demand and may be destined to effect. In a word, and
speaking broadly, we wish to know not merely in a

narrow sense what the new idea is, but, in the larger meaning of the term, what it " can."

I shall first speak briefly of the so-called " positive " character of the definition, an alleged essential quality of it, a seeming property which criticism is wont to signalise as a radical or intrinsic virtue of the concept itself. Quite independently of the mathematicians Dedekind and Cantor, who, we have seen, were the independent originators of the new formulation, the then old philosopher, Bolzano, bringing to the subject another order of training and of motive, arrived at notions of the finite and infinite, which on critical examination are found to be essentially the same as theirs, though greatly differing in point alike of view and of form. Bolzano's procedure is virtually as follows: — Suppose given a class C of elements, or things, of any kind whatsoever, as the sands of the seashore, or the stars of the firmament, or the points of space, or the instants in a stretch of time, or the numbers with which we count, or the total manifold of truths known to an omniscient God. Out of any such class C, suppose a series formed by taking for first term one of the elements of C, for second term two of them, and so on. Any term so obtainable is itself obviously a class or group of things, and is *defined* to be finite. The indicated process of series formation, if sufficiently prolonged, will either exhaust C or it will not. If it will, C is itself *demonstrably* finite; if it will not, C is, on that account, *defined* to be infinite. Now, say Professor Royce and others, a definition like the latter, being dependent on such a notion as that of inexhaustibility or endlessness or boundlessness, is negative; a certain innate craving of the understanding remains unsatisfied, we are told, because the definition presents the notion, not in a positive way by telling us what the infinite actually *is*, but merely in a negative fashion by telling us what it is *not*. Undoubtedly the claim is plausible, but is it more? Bolzano affirmed and exemplified a certain proposition, in itself of the utmost importance, and throwing half the needed light upon the question in hand. That proposition is: *Any class or assemblage (of elements), if infinite according to his own definition of the term, enjoys the property of being equivalent, in the sense above explained, to some proper part of itself.* Though he did not himself demonstrate the proposition, it readily admits of demonstration, and, since his time, has in fact been repeatedly and rigorously proved. Not only that, but the converse proposition, giving the other half of the needed light, has been established too: *Every assemblage that* HAS *a part " equivalent " to the whole,* IS *infinite in the Bolzano sense of the term.*

It so appears, in the conjoint light of those two theorems, that the property seized upon and pointed out by the ingenious theologian is in all strictness a *characteristic,* though derivative, mark of the infinite as he conceived and defined it. It is sufficiently obvious, therefore, that this derivative property might logically be regarded as *primitive,* made to serve, that is, as a ground of definition. Precisely this fact it is which was independently perceived by Dedekind and Cantor, with the result that, as they have presented the matter, a collection, or manifold, is infinite if it *has* a certain property, and finite if it has it *not.* And now, the critics tell us, it is the infinite which is positive and the finite which is negative.

The distinction appears to me to be entirely devoid of essential merit. It seems rather to be only another interesting example of that verbal legerdemain for which a certain familiar sort of philosophising has long been

famous.   For what indeed is positive and what negative?
Are we to understand that these terms have absolute
as distinguished from relative meaning?   The distinction,
I take it, is without external validity, is entirely subjective, a matter quite at will, being dependent solely on an
arbitrary *ordering* of our thought.   That which is first
put in thought is positive: the opposite, being subsequently put, is negative; but the *sens* of the time-vector
joining the two may be reversed at the thinker's will.   It
is sometimes contended that that which *generally* happens
in the world, and so constitutes the *rule,* is intrinsically
positive.   As a matter of fact a moving body " in
general " continuously changes its distance from every
object.   Such change of distance from *every* other object
would accordingly be a positive something.   Then it
would follow that the classic definition of a sphere-surface
as the locus of a moving point which does *not* change its
distance from a certain specified point, is really negative.   Obviously it avails nothing essential to disguise the
negativity by some such seemingly positive phrase as
" constant " distance.   The trick is an easy one.   If,
again, it be allowed that, a process being once started, its
continuation is positive, its termination negative, then it
would result that *in*-exhaustibility is positive and ex-
haustibility negative, whence we should have to own that
it is Bolzano's definition which is positive and that by
Dedekind and Cantor negative.   It hardly admits of
doubt that the matter is purely one of an arbitrarily
chosen point of view.   The distinction is here of no importance.   What is important is that, no matter which
of the definitions be adopted as such, the other then
states a derivable property of the thing defined.   In either
case the *concept* of the infinite remains the same, it is
merely its *garb* that is changed.   I am very far from

intending, however, to assert herewith that, because the definitions are logically equivalent, they must needs be or indeed are so practically, that is, as instruments of investigation. That is another matter, which, I regret to say, our somewhat pretentious critiques of scientific method furnish no better means of settling than the wasteful way of trial. Everyone will recall from his school-days Euclid's definition of a plane as being a surface such that a line joining any two points of the surface lies wholly in the surface. Logically that is equivalent to saying: A plane is such an assemblage of points that, any three independent points of the assemblage being given, one and only one third point of the assemblage can be found which is equidistant from the given three. But, despite their logical equivalence, who would contend that, for elementary purposes, the latter notion is " practically " as good as the Greek? And so in respect to the infinite, I am free to admit, or rather I affirm, that, on the score of usability, the Dedekind-Cantor definition is greatly superior to its Bolzanoan equivalent. Professor Royce has indeed ingeniously shown how readily it lends itself to philosophic and even to theologic uses.

I turn now to the current assertion by Professor Royce and Mr. Russell, that the modern concept of the infinite, of which I have given above in italics an exact statement, to which the reader is referred, in fact denies a certain ancient axiom of common sense, namely, the axiom of whole and part. I am not about to submit a brief in behalf of the traditional conception of axioms as self-evident truths. That conception, as is well known, has been once for all abandoned by philosophy and science alike, while to mathematicians in particular no phenomenon is more familiar than that of the co-existence of self-coherent bodies of doctrine constructed on distinct and self-consistent but incompatible systems of postu-lates. The co-ordination of such incompatible theories is quite legitimate and presents no cause for regret or alarm. The forced recession of the axioms from the high ground of absolute authority, so far from indicating chaos of intellection or ultimate dissolution of knowledge, signifies a corresponding deepening of foundation; it means an ascension of mind, the proclamation of its creative power, the assertion of its own supremacy. And henceforth the denial of specific axioms, or the deliberate substitution of one set for another, is to be rightly re-garded as an inalienable prerogative of a liberated spirit. The question before us, then, is one merely of fact, namely, whether a certain axiom is indeed denied or contradicted by the modern concept of the infinite.

It is in the first place to be observed that the state-ment itself of that concept avoids the expression, " equal-ity of whole and part," but instead of it deliberately employs the term " equivalence." The word actually used by Dedekind himself is *ähnlichkeit* (similarity). But, says Professor Royce, " equivalence " is just what the axiom really means by equality. It is precisely this statement which I venture to draw in question. If we know that each soldier of a company marching along the street has one and but one gun on his shoulder, then, we are told, even if we do not know *how* many soldiers or guns there are, we do know that there are " *as* many " soldiers as guns. What the definition in question, taken severely, itself affirms in this case, is that the assemblage of guns is " equivalent or similar " to that of the soldiers. Let us now suppose that in place of the soldiers we write, for example, " all positive integers," and in place of guns, " all even positive integers " — the integers are plainly susceptible of unique and reciprocal association with the even integers, — then the definition again asserts, as before, " equivalence " of these assemblages. Note that thus far nothing has been said about *number* as an expression of *how many.* If there *be* a number that tells how many things there are in one assemblage, that same number doubtless tells how many there are in any " equivalent " assemblage, and just because the number, if there be one, is the *same* for both, the two are said to be *equal* by axiom. In this view, equality of groups means more than mere " equivalence "; it means, besides, sameness of their numbers, and so applies *only* in case there be numbers. But common sense, whose axiom is here in court, has neither found, nor affirmed the existence of, a number telling, for example, how many integers there are. On the other hand, in case of assemblages for which common sense *has* known a number, the axiom of whole and part is admittedly valid without exception. It thus appears that the axiom supposed, regarded, however unconsciously but nevertheless in intention, as applicable only in case there be a number telling how many, is, in all strictness, not denied by the concept in question. Numbers designed to tell how many elements there are in an assemblage having a part " equivalent " to the whole are of recent invention, and it may be remarked in passing that this invention bears immediate favourable witness to the fruitfulness of the new idea. Such transfinite numbers once created, then undoubtedly, and not before, the question naturally presents itself whether " equivalence " shall be translated " equality," or, what is tantamount, whether the latter term shall be generalised into the former; " generalised," I say, for, though it is true that, as soon as the transfinite numbers are created, there is, in case of an infinite col-

lection and some of its parts, a conjunction of " equivalence " and " sameness of number," yet equality does not
of itself deductively attach, for the transfinite numbers
are in *genetic* principle,[8] *i.e.*, radically, different from the
number notion which the concept of equality has hitherto connoted. The question as to the mentioned translation or generalisation is, therefore, a question, and it
is to be decided, not under spur or stress of logic, but
solely from motives of economy acting on grounds of
pure expedience. If the decision be, as seems likely
because of its expedience and economy favourable to such
translation or generalisation, then indeed the old axiom,
as above construed, still remains uncontradicted, is yet
valid within the domain of its asserted validity. It is
merely that a new number-domain has been adjoined
which the old verity never contemplated, and in which,
therefore, though it does not apply, it never essentially
pretended to; but on account of which adjunction, nevertheless, for the sake of good neighbourship, it is constrained, not indeed to retract its ancient claims, but
merely to assert them more cautiously and diplomatically,
in preciser terms. Even then, in case of quarrel, it is the
generaliser who should explain, and not a defender of
the generalised.

And now to my final thesis I venture to invite the
reader's special attention, and beg to be held with utmost
strictness accountable for my words. The question is,
whether it is possible, by means of the new concept, to
demonstrate the existence of the infinite; whether, in
other words, it can be proved that there are infinite systems. That such demonstration is possible is affirmed
by Bolzano, by Dedekind, by Professor Royce, by Mr.
Russell, and in fact by a large and swelling chorus of

---

[8] *Cf.* Couturat, *L'Infini mathématique,* Appendix.

authoritative utterance, scarcely relieved by a dissenting voice. After no little pondering of the matter, I have been forced, and that, too, I must own, against my hope and will, to the opposite conviction. Candour, then, compels me to assert, as I have elsewhere [9] briefly done, not only that the arguments which have been actually adduced are all of them vitiated by circularity, but that, in the very nature of conception and inference, by virtue of the most certain standards of logic itself, every potential argument, every possible attempt to prove the proposition, is foredoomed to failure, destined before its birth to take the fatal figure of the wheel.

The alleged demonstrations are essentially the same, being all of them but variants under a single type. It is needless, therefore, in support of my first contention, to present separate examination of them all. Analysis of one or two specimens will suffice. I will begin with one from Bolzano's offering, both because it marks the beginning of the new era of thought about the subject and because subsequent writers have nearly all of them either cited or quoted it, and that, as far as I am aware, always with approval. Bolzano [10] undertakes to demonstrate, among similar statements, the proposition that *die Menge der Sätze und Wahrheiten an sich* is infinite (*unendlich*), this latter term being understood, of course, in accordance with his own definition above given. The attempt, as anyone may find who is willing to examine it minutely, informally postulates as follows: the proposition, There are such truths (as those contemplated in the proposition), is such a truth, *T; T* is true, is another such truth, *T;* so on; *and,* the indicated process is inexhaust-

---

[9] "The Axiom of Infinity and Mathematical Induction," *Bulletin of the American Mathematical Society,* vol. ix., May, 1903.

[10] "Paradoxien," sect. 14.

ible. Now, these assumptions, which are essential to the argument, and which any careful reader cannot fail to find implicit in it, are, possibly, all of them, correct, but the last is so evident a *petitio principii* as to make one look again and again lest his own thought should have played him a trick.

In case of Dedekind's demonstration, which has been heralded far and wide, the fallacy is less glaring. The argument is far subtler, more complicate, and the *versteckter Zirkel* lies deeper in the folds. But it is undoubtedly there, and its presence may be disclosed by careful explication. Let the symbol t stand for thought, *any* thought, and denote by t' the thought that t *is* a thought. For convenience, t' may be called the image of t. On examination, Dedekind's proof is found to *postulate* as *certainties:* (1) If there be a t, there is a t', image of t; (2) if there be two distinct t's, the corresponding t''s are distinct; (3) there is a t; (4) there is a t which is not a t'; (5) every t is *other* than its t'. These being granted, it is easy to see, by supposing each t to be paired with its t', as object with image, that the assemblage θ of all the t's and the assemblage θ' of all t''s are " equivalent." But by (4) there is a t not in θ' which latter is, therefore, a *part* of θ. Hence θ is infinite, by definition of the term.

Let this matter be scrutinised a little. Assuming only the mentioned postulates and, of course, the possibility of reflection, it is obvious that by pairing the t of (4) with its image t', then the latter with *its* image, and so on a sequence S of t's is started which, because of (1) and (5), is incapable of termination. This S, too, by Dedekind's proof, is an infinite assemblage. Accordingly, postulate (1), without which, be it observed, the proof is impossible, postulates, in *advance* of the argument, cer-

tainty which, if the argument's conclusion be true, *tran-scends* the *finite before* the inference that an *infinite* exists either is or can be *drawn*. The reader may recall how the Russian mathematician Lobatschewsky said, " In the absence of proof of the Euclidian postulate of parallels, I will assume that it is not true "; and how thereupon there arose a new science of space. Suppose that, in like manner, we say here, " In the absence of *proof* that an act once found to be mentally performable is endlessly so performable, we will assume that such is not the case," then, whatever else might result — and of that we shall presently speak — one thing is at once absolutely certain: Dedekind's " argument " would be quite impossible. The fact is that a more beautiful circle than his is hardly to be found in the pages of' fallacious speculation, or admits of construction by the subtlest instruments of self-deceiving dialectic, though it must be frankly allowed that Mr. Russell's [11] more recent movement about the same centre is equally round and exquisite.

And this disclosure of the fatal circle in the attempted demonstration serves at once to introduce and exemplify the truth of my second contention, which is that all logical discourse, of necessity, *ex vi termini,* presupposes certainty that transcends the finite, where by logical discourse I mean such as consists of completely determined concepts welded into a concatenated system by the ancient hammer of deductive logic. The fact of this presupposition, of course, cannot be *proved,* but, and that is good enough, it can be *exhibited* and beheld. To attempt to " prove " it would be to stultify oneself by assuming the possibility of a deductive argument A to prove that the conclusion of A cannot be drawn unless

[11] *Principles of Mathematics,* chap. xliii.

it is assumed in advance. The fact, then, if it be a fact, and of that there need not be the slightest doubt, is to be added to that small group of fundamental simplicities which can at best be *seen,* if the eye be fit.

Consider, for example, this simplest of syllogistic forms: Every element e of the class c is an element e' of the class c'; every e' of c' is an element e'' of the class c''; $\therefore$ every e of c is an e'' of c''. I appeal now to the reader's own subjective experience to witness to the following facts: (1) Our *apodictic feeling* is the sole justification of the inference as such; (2) that felt justification is absolute, neither seeking nor admitting of appeal; (3) that sole and absolute justification, namely, the apodictic feeling, is in no slightest degree *contingent* upon the answer to any question whether the multitude of elements e or e' or e'' is or is not, may or may not be found to be, "equivalent" to some part of itself. The feeling of validity here undoubtedly transcends the finite, undoubtedly holds naught in reserve against any possibility of the inference failing as an act should the system of elements turn out to be infinite.

At some risk of excessive clearness and accentuation, for the matter is immeasurably important, I venture to ask the reader to witness how the transcendence or transfiniteness of certainty shows itself in yet another way, not merely in formal deductive *inference,* but also in *conception.* When any concept, as that of Parabola, for example, is formed or defined, it is found that the concept contains implicitly a host of properties not given explicitly in the definition. Properly speaking, the thing defined *is* a certain organic assemblage of properties, of which the totality is implied in a properly selected few of them. Now the fact which it is decisive here to note is that by conception we mean, among other things, that

whenever the definition may present itself, even though
it may be endlessly, a certain invariant assemblage of
properties implicitly accompanies the presentation. Without such transfinite certainty of such invariant uncon-
tingent implication, conception would be devoid of its
meaning.

The upshot, then, is this: that conception and logical
inference alike presuppose absolute certainty that an act
which the mind finds itself capable of performing is intrinsically performable endlessly, or, what is the same
thing, that the assemblage of possible repetitions of a
once mentally performable act is equivalent to some
proper part of the assemblage. This certainty I name
the *Axiom* of *Infinity,* and this axiom being, as seen, a
necessary presupposition of both conception and deductive inference, every attempt to " demonstrate " the existence of the infinite is a predestined begging of the issue.

What follows? Do we, then, *know* by axiom that the
infinite is? That depends upon your metaphysic. If
you are a radical *a-priorist,* yes; if not, no. If the latter,
and I am now speaking as an a-priorist, then you are
agnostic in the deepest sense, being capable, in utmost
rigour of the terms, of neither conceiving nor inferring.
But if we do not *know* the axiom to be true, and so
cannot deductively prove the existence of the infinite,
what, then, is the *probability* of such existence? The
highest yet attained. Why? Because the *inductive* test
of the axiom, regarded now as a hypothesis, is trying to
conceive and trying to infer, and this experiment, which
has been world-wide for æons, has seemed to succeed in
countless cases, and to fail in none not explainable on
grounds consistent with the retention of the hypothesis.

Finally, to make briefest application to a single concrete case. Do the stars constitute an infinite multitude?

No one knows. If the number be finite, that fact may some time be ascertained by actual enumeration, and, if and only if there be infinite ensembles of possible repetitions of mental processes, it may also be known by proof. But if the multitude of stars be infinite, that can never be known *except* by proof; this last is possible only if the axiom of infinity be true, and even if this be true, the actual proof may never be achieved.

THE PERMANENT BASIS OF A LIBERAL EDUCATION [1]

Is it possible to find a principle or a set of principles qualified to serve as a permanent basis for a theory of liberal education? If so, what is the principle or set of principles? These are old questions. We are living in a time when they must be considered anew.

If our world were a static affair, if our environment, physical, spiritual and institutional, were stable, then we should none of us have difficulty in agreeing that a liberal education would be one that gave the student adjustment and orientation in the world through disciplining his faculties in their relation to its cardinal static facts. Such a world could be counted upon. No one doubts that in such a world it would be possible to find a permanent basis for a theory of liberal education — a principle or a set of principles that would be adequate and sound, not merely to-day, but to-day, yesterday and to-morrow.

But we are reminded by certain rather numerous educational philosophers that our world is not a static affair. We are told that it is a scene of perpetual change, of endless and universal transformation — physical flux, institutional flux, social flux, spiritual flux: all is flux. These philosophers tell us of the rapid and continued advancement and multiplication of knowledge. They do not cease to remind us that knowledge goes on building itself out, not only in all the old directions, but also in

[1] Printed in *The Columbia University Quarterly*, June, 1916.

an endlessly increasing variety of new directions. They remind us that the ever-augmenting volume of knowledge is continually breaking up into new divisions or kinds, and that each of these quickly asserts, and sooner or later demonstrates, its parity with any other division in respect of utility and dignity and disciplinary value. They remind us that a striking concomitant phenomenon, which is partly the effect of the multiplication and differentiation of knowledge, partly a cause of it and partly owing to other agencies and influences, is the fact that new occupations constantly spring into being on every hand and that the needs, the desires and the habits of men, and therewith the drifts and forms of social and institutional life, suffer perpetual mutation. Nothing, they tell us, is permanent except change itself. All things, material, mental, moral, social, institutional, are tossed in an infinite and endless welter of transformations — evolution, involution, revolution, all going on at once and forever.

It is evident, we are assured, that in such a world the search for abiding principles is vain, whether we seek a permanent basis for a liberal education or a permanent basis for anything else. The doctrine is that in our world permanent bases do not exist. Permanence, stability, invariance, immutability, there is none. It exists only in rationalistic dreams. It exists only in the insubstantial musings of the tender-minded. It exists only in the cravings of such as have not the pragmatistic courage or constitution to deal with reality as it is in the welter and the raw. We are told that there is in matters educational no such thing as eternal wisdom. Wisdom is at best a transitory thing, depending on time and place, and constantly changing with them. A prescription that is wise to-day will be foolish to-morrow. What *was* a liberal

education is not such now. What is a liberal education
to-day will not be liberal in the future. Greek has gone,
theology is gone, religion is gone, Latin is almost gone,
mathematics, we are told, is going, and so on and on.
Each branch of knowledge will have its day, and then
will cease to be essential. Liberal curricula, it is contended, must change with the times.

This doctrine, logically conceived and carried out,
means that as the years and generations follow endlessly,
time and change will beget an endless succession of so-
called liberal curricula. It means that, if, in this un-
ending sequence, we observe a finite number of succes-
sive curricula, these will indeed be found to resemble each
other, overlapping, interpenetrating, and thus seeming
to be held together in a kind of unity by a more or less
vague and elusive bond; but that this must be appearance only. For if the observed succession be prolonged,
as it is bound to be, the seeming principle of unity must
become dimmer and dimmer; the terms or curricula of
the endless succession of them can have, in fact, nothing
in common, no *lien*, no unity whatever, save that pale
variety which serves merely to constitute the succession
of curricula an infinite series of terms. It is not unlikely
that the educational philosophers in question may not be
aware that this is what their doctrine means. Nevertheless, that is what it does mean.

Is the doctrine sound? To me it seems not to be so.
The question is a question of fact. The denial of permanent principle and the assertion of its concomitant
theory of education seek to justify themselves by pointing to the fluctuance of the world. I do not deny the
fluctuance of the world. One must be blind to do that.
Here, there and yonder, in the world of matter, in the
world of mind, in thought, in religion, in morals, in con-

ventions, in institutions, everywhere are evident the drift-
ings and shiftings of events: everywhere course the hast-
ing streams of change. I admit the storm and stress,
the tumult and hurly-burly of it all. I do not deny that
impermanence is a permanent and mighty fact in our
world. What I do deny is that impermanence is universal. Its sweep is not clean. Far from it. If it is,
man has indeed been a colossal fool, for the quest of
constance, the search for invariance, for things that
abide, for forms of reality that are eternal, has been in
all times and places the dominant concern of man, uniting his philosophy, his religion, his science, his art and
his jurisprudence into one manifold enterprise of mankind. Not permanence alone, nor impermanence alone,
but the two together, one of them drawing and the other
driving, it is these *two* working together that have shaped
the course of human history and moulded the form of its
content. I admit that impermanence is more evident and
obtrusive than permanence, but I contend that a philosophy which finds in the world nothing but change is a
shallow philosophy and false. The instinct that perpetually drives man to seek the fixed, the stable, the
everlasting, has its root deep in nature. It is a cosmic
thing. Must we say that this instinct, this most imperious
of human cravings, has no function except that of quali-
fying man to be eternally mocked? It cannot be admitted. The sweep of mutation is indeed deep and wide,
but it is not universal. It would be possible, in a contest
before a committee of competent judges, to show that
temporalities are, in respect of number, more than
matched by eternalities, and that, in respect of relative
importance, changes are as dancing wavelets on an infinite and everlasting sea.

In our environment there exist certain great invariant

massive facts that now are and always will be necessary and sufficient to constitute the basis of a curriculum or a theory of liberal education. These facts are obvious and on that account they require to be pointed out, just because, in the matter of escaping attention, what is very obvious is a rival of what is obscure.

What are these facts? One of them is the fact that every human being has behind him an immense human past, the past of mankind. Of course, I do not mean that what we call the human past is itself a fixed or permanent thing. It is not. It is a variable, constantly changing by virtue of perpetual additions to it as the years and centuries empty the volume of their events into that limitless sea. What is permanent is the fact — it was so yesterday and it will be so to-morrow — that behind each one of us there is a human past so immense as to be practically infinite. That fact, I say, is permanent. It can be counted on. It is as nearly eternal as the race of man. Out of that past we have come. Into it we are constantly passing. Meanwhile, it is of the utmost importance to our lives. It contains the roots of all we are, and of all we have of wisdom, of science, of philosophy, of art, of jurisprudence, of customs and institutions. It contains the record or ruins of all the experiments that man has made during a quarter or a half million years in the art of living in this world. This great stable fact of an immense human past behind every human being that now is or is to be, obviously makes it necessary for any theory of liberal education to provide for discipline in human history and in the literature of antiquity. How much? A reasonable amount — enough, that is, to orient the student in relation to the past, to give him a fair sense of the continuity of the life of mankind, a decent appreciation of ancient works of genius, and sense and knowledge enough to guide his energies and to control his enthusiasms in the light of human experience. As the centuries go by, ancient literature and human history will increase more and more. What is a reasonable prescription will, therefore, become less and less in its relation to the increasing whole, but it will never vanish. It will never cease to be indispensable.

In this connection, the following question is certain to be asked. From the point of view of this inquiry, which aims at indicating an enduring basis for a theory of liberal education, does it follow that Greek or Latin or any other language that may be destined to become "classic" and "dead" at some remote future time, — does it follow that these or any of them must enter as essential into the curriculum of a liberal education? It does not. It would indeed be a grave misfortune if there should ever come a time when there were no longer a goodly number of scholars devoted to the great languages of antiquity. Some of the thought, of the science, of the wisdom, of the beauty originally expressed in these tongues, is, we have said, essential; but it is precisely the chief function of those who master the ancient languages to make their precious content available, through translations and critical commentaries, for the great body of their fellow men to whom the languages themselves must remain unknown. It is not denied that the scholars in question will know and appreciate such content as no others can, but neither will these scholars continue forever to deny the possibility of rendering most of the content *reasonably* well in the living languages of their fellow men. The contrary cannot be much longer maintained. Indeed the layman already knows that Euclid, Plato, Aristotle, Aeschylus, Sophocles, Euripides, Demosthenes, Virgil, Cicero, Lucretius, and many others, have

already learned, or are rapidly learning, to speak, beautifully and powerfully, all the culture languages of the
modern world.

Another of the massive facts that transcends the flux
of the world, and that, therefore, must contribute basic-
ally to any permanent theory of liberal education, is the
fact that every human being is encompassed by a physical or material universe. Again I do not mean, of course
I do not mean, that the universe remains always the
same. What is permanent is the fact that human beings
always have been, now are, and always will be, sur-
rounded on every hand by an infinite objective world of
matter and force. In that world we are literally immersed. Our bodies are parts of it; they are composed
of its elements and will be resolved into them again. If
our minds, too, be not a part of it, they must at all events,
on pain of our physical incompetence or extinction, gain
and maintain continuous and intelligent relations with it.
The great fact in question, like the fact of the human
past, can be counted on. It survives all vicissitudes.
The immersing universe may be a chaos or a cosmos, or
partly chaotic and partly cosmic, preserving its character
in that regard or tending along an asymptotic path to
chaos complete or to cosmic perfection. But if it *is*
chaotic, we humans sufficiently match it in that regard
to be able to treat it more and more successfully as if it
were an infinite locus of order and law. And we know
that to do this is immensely advantageous. In a strict
sense, it is absolutely indispensable. Merely to live, it is
necessary to treat nature as having *some* order.

These considerations show that any theory which aims
to orient and discipline the faculties of men and women
in their relation to the great permanent facts of the world
must make basal provision for discipline in what we call

natural, or physical, science. Again, how much? Again the answer is, a reasonable amount. But how much, pray, is that? Enough to give the student a fair acquaintance with the heroes of natural science, a fair understanding of what scientific men mean by natural order or law, a decent insight into scientific method, the rôle of hypothesis, and the processes of experimentation and verification. But there are so many branches of natural science and their number is increasing. A liberal curriculum cannot require them all. Which shall be chosen? It does not matter much. These branches differ a good deal in content and in a less degree in method, but they have enough in common to make a claim of superiority for any one of them mainly a partisan claim. The spirit of science, its methods, some of its chief results, these are the essentials. To give these, physics is competent, so is chemistry, so is botany, so is zoology, and so on. The choice is a temporal detail, but the principle requiring the choice is everlasting. A hundred or a thousand years hence, there will be other details to choose from — scientific branches not yet named, nor even dreamed of. But — and this is the point — a theory of liberal education will not cease to demand some discipline in natural science so long as human beings are immersed in an infinite world of matter and force.

Nor will such a theory fail to take account fundamentally of a third great fact that persists despite the flux of things and the law of death. I refer to the fact that every human being's fortune depends vitally upon what may be called the world of ideas. It is evident that of the total environment of man, the human *Gedankenwelt* is a stupendous and mighty component. Like the other great components already named, or namable, the world of ideas is, in respect of its existence, a permanent datum amid the weltering sea of change. Not only may it be counted on, but it must be reckoned with. Some thinking everyone must do. The formation and combination of ideas is not merely indispensable to welfare, it is more fundamental than that: it is essential to human life. The world of ideas contains countless possibilities that are not actualized or realized or validated or incarnated, as we say, in the order of the material world, nor in any existing social or institutional order. It is plain that discipline in the ways and forms of abstract thinking, of dealing with ideas as ideas, is essential to a liberal education, not merely because the world of ideas is itself a thing of supreme and eternal worth, but because those who are incapable of constructing ideal orders may not hope to have the imagination requisite for ascertaining or for appreciating the frame and order actualized in external nature. From all of this it is clear that any enduring theory of liberal education must provide for the discipline of logic and mathematics, for it is in these and these alone that rigorous or cogent thinking finds its standard and its realization. It is true that most of the thinking that the exigencies of life compel us to do is not cogent thinking. We are obliged constantly to deal with ideas that are too nebulous to admit of rigorously logical handling. But to argue that consequently discipline in rigorous thinking is not essential, is stupid. It is to ignore the value of standards and ideals. It is, in other words, to be spiritually blind. I am making no partisan plea for my own subject. Mathematics happens to be the name that time has given to rigorous or cogent thinking, and so it happens that mathematics is the name of the one art or science that is qualified to give men and women a perfect standard of thinking and to bring them into the thrilling presence of indestructible bodies of thought. Call the science by any other name — anathematics or logostetics. The thing itself and its functions would be the same.

Another cardinal fact among the permanent considerations that a theory of liberal education must rest upon is the fact that human beings are social beings. It is only in dreams and romances that a human being lives apart in isolation. Men, said Aristotle, are made for co-operation. Every man and every woman is a born member of a thousand teams. Not one is pure individual. Each one is many. None can extricate himself from the generic web of man. This fact survives the flux. It is as nearly everlasting as the human race. It is a rock to build upon. And so it was true yesterday, is true to-day, and will be true to-morrow, that an education whose function it is to discipline the faculties of man in their relation to the great abiding facts of life and the world, must provide for discipline in the fundamentals of political science. Moreover, as it is essential to the health and to the effectiveness of the individual, and also essential to the welfare of society that men and women be able to express themselves acceptably and effectively, a liberal education will provide for discipline in the greatest of all the arts — the art of rhetoric. No term has been more abused, especially by amorphous men of science. Yet the late Henri Poincaré was made a member of the French Academy, not because he was a great mathematician, astronomer, physicist and philosopher, but because of his masterful control of the resources of the French language as an instrument of human expression.

I have spoken of the invariant fact of the human past. Its complement is the fact of the human future. That, too, is a great abiding fact. It is, in practice, to be treated as eternal, for, if the race of man be doomed to extinction, then, in that far-off event, human education itself will cease. Does it follow that a theory of liberal education must provide for instruction in prophecy? It does follow. But is it not foolish to speak of instruction in prophecy? For is not prophecy a thing of the past? Is it not a dead or a dying office of priests? It is not foolish, it is not a thing of the past, it is not a dead or dying office of priests. Prophecy is a thing of the present, destined to increase with the advancement of knowledge. Every department of study is a department of prophecy. It is the function of science to foretell. Prophecy is not the opposite of history, it is history's main function. As W. K. Clifford long ago pointed out, every proposition in physics or astronomy or chemistry or zoology or mathematics, or other branch of science, is a rule of conduct facing the future — a rule saying that, if such-and-such be true, then such-and-such must be true; if such-and-such a situation be present, then such-and-such things will happen; if we do thus-and-thus, then certain statable consequences may be expected. Foretelling, indeed, is not the exclusive office of knowledge, for musing, meditation, pensiveness, pure contemplation, have their legitimate place; but man is mainly and primarily an active being; and in relation to action, the business of knowledge is prophecy, forecasting what to do and what to expect.

Finally it remains to mention another fundamental matter that must contribute in a paramount measure to any just theory of a liberal education. It is not a matter strictly co-ordinate with the other matters mentioned, but it touches them all, penetrates them all and transfigures them all. I refer to the discipline of beauty. Beauty is the most vitalizing thing in the world. It is beauty that makes life worth living and makes it pos-

sible. If, by some fiendish cataclysm, all the beauty of
art and all the beauty of nature were to be suddenly
blotted out, the human race would quickly perish through
depression caused by the ubiquitous presence of ugliness.
Does it follow that a liberal curriculum must provide for
the instruction of every student in all the arts? No.
Like the natural sciences, the arts are enough alike to
make any one of them a representative of them all.
Besides, all subjects of study are penetrated with beauty,
and any one of them may be so administered as to enlarge and refine the sense of what is beautiful in life
and the world.

Such I take to be major considerations among the
great permanent massive facts that together suffice and
are essential to constitute an enduring basis for a theory
of liberal education. Ought discipline to be prescribed
in all the indicated fields? The answer would seem to be
that a liberally educated man or woman is one who has
been instructed in them all. It follows that there be
seekers who are by nature not qualified to find. But in
the case of these, as in the case of their more gifted fellows, it must be remembered that not the least service a
program of liberal study should render, is that of disclosing to men and women and to their fellows their
respective powers and limitations.

GRADUATE MATHEMATICAL INSTRUCTION FOR GRADUATE STUDENTS NOT INTENDING TO BECOME MATHEMATICIANS [1]

In his "Annual Report" under date of November last, the President of Columbia University speaks in vigorous terms of what he believes to be the increasing failure of present-day advanced instruction to fulfil one of the chief purposes for which institutions of higher learning are established and maintained.

In the course of an interesting section devoted to college and university teaching, President Butler says:

A matter that is closely related to poor teaching is found in the growing tendency of colleges and universities to vocationalize all their instruction. A given department will plan all its courses of instruction solely from the point of view of the student who is going to specialize in that field. It is increasingly difficult for those who have the very proper desire to gain some real knowledge of a given topic without intending to become specialists in it. A university department is not well organized and is not doing its duty until it establishes and maintains at least one strong substantial university course designed primarily for students of maturity and power, which course will be an end in itself and will present to those who take it a general view of the subject-matter of a designated field of knowledge, its methods, its literature and its results. It should be possible for an advanced student specializing in some other field to gain a general knowledge of physical problems and processes without becoming a physicist; or a general knowledge of chemical problems and processes without becoming a chemist; or a general knowledge of zoological problems and processes without becoming a zoologist; or a general knowledge of mathematical problems and processes without becoming a mathematician.

[1] An address delivered before Section A of the American Association for the Advancement of Science, December 30, 1914. Printed in *Science,* March 26, 1915.

This is a large matter, involving all the cardinal divisions of knowledge. I have neither time nor competence to deal with it fully or explicitly in all its bearings. As indicated by the title of this address it is my intention to confine myself, not indeed exclusively but in the main, to consideration of the question in its relation to advanced instruction in mathematics. The obvious advantages of this restriction will not, I believe, be counterbalanced by equal disadvantages. For, much as the principal subjects of university instruction differ among themselves, it is yet true that as instruments of education they have a common character and for their efficacy as such depend fundamentally upon the same educational principles. A discussion, therefore, of an important and representative part of the general question will naturally derive no little of whatever interest and value it may have from its implicit bearing upon the whole. It is not indeed my intention to depend solely upon such implicit bearings nor upon the representative character of mathematics to intimate my opinion respecting the question in its relation to other subjects. On the contrary, I am going to assume that specialists in other fields will allow· me, as a lay neighbor fairly inclined to minding his own affairs, the privilege of some quite explicit preliminary remarks upon the larger question.

I suspect that my interest in the matter is in a measure temperamental; and my conviction in the premises, though it is not, I believe, an unreasoned one, may be somewhat colored by inborn predilection. At all events I own that a good many years of devotion to one field of knowledge has not destroyed in me a certain fondness for avocational studies, for books that deal with large subjects in large ways, and for men who, uniting the generalist with the specialist in a single gigantic personality, can show you perspectives, contours and reliefs, a great subject or a great doctrine in its principal aspects, in its continental bearings, without first compelling you to survey it pebble by pebble and inch by inch. I can not remember the time when it did not seem to me to be the very first obligation of universities to cherish instruction of the kind that is given and received in the avocational as distinguished from the vocational spirit — the kind of instruction that has for its aim, not action but understanding, not utilities but ideas, not efficiency but enlightenment, not prosperity but magnanimity. For without intelligence and magnanimity — without light and soul — no form of being can be noble and every species of conduct is but a kind of blundering in the night. I could hardly say more explicitly that I agree heartily and entirely with the main contention of President Butler's pronouncement. Indeed I should go a step further than he has gone. He has said that a university *department* is not well organized and is not doing its duty until it establishes and maintains the kind of instruction I have tried to characterize. To that statement I venture to add explicitly — what is of course implicit in it — that a *university* is not well organized and is not doing *its* duty until it makes provision whereby the various departments are enabled to foster the kind of instruction we are talking about. That in all major subjects of university instruction there ought to be given courses designed for students of " maturity and power " who, whilst specializing in one subject or one field, desire to generalize in others, appears to me to be from every point of view so reasonable and just a proposition that it would not occur to me to regard it as questionable or debatable were it not for the fact that it actually is questioned and debated by teachers of eminence and authority.

What is there in the contention about which men may differ? Dr. Butler has said that there is a " growing tendency of college and university departments to vo-cationalize all their instruction." Is the statement erroneous? It may, I think, be questioned whether the tendency is growing. I hope it is not. Of course specialization is not a new thing in the world. It is far older than history. Let it be granted that it is here to stay, for it is indispensable to the advance-ment of knowledge and to the conduct of human affairs. Every one knows that. There is, however, some evi-dence that specialization is becoming, indeed that it has become, wiser, less exclusive, more temperate. The symptoms of what not long ago promised to become a kind of specialism mania appear to be somewhat less pronounced. Recognition of the fact that specializa-tion is in constant peril of becoming so minute and narrow as to defeat its own ends is now a commonplace among specialists themselves, many of whom have learned the lesson through sad experience, others from observation. Specialists are discoverers. One of our recent discoveries is the discovery of a very old truth: we have discovered that no work can be really great which does not contain some element or touch of the universal, and that is not exactly a new insight. Leo-nardo da Vinci says:

We may frankly admit that certain people deceive themselves who apply the title " a good master " to a painter who can only do the head or the figure well. Surely it is no great achievement if by studying one thing only during his whole lifetime he attain to some degree of excel-lence therein!

The conviction seems to be gaining ground that in the republic of learning the ideal citizen is neither the ignorant specialist, however profound he may be, nor the shallow generalist, however wide the range of his interest and enlightenment. It is not important, however, in this connection to ascertain whether the vocationalizing tendency is at present increasing or decreasing or stationary. What is important is to recognize the fact that the tendency, be it waxing or waning, actually exists, and that it operates in such strength as practically to exclude all provision for the student who, if I may so express it, would qualify himself to gaze into the heavens intelligently without having to pursue courses designed for none but such as would emulate a Newton or a Laplace. If any one doubts that such is the actual state of the case, the remedy is very simple: let him choose at random a dozen or a score of the principal universities and examine their bulletins of instruction in the major fields of knowledge.

Another element — an extremely important element — of President Butler's contention is present in the form of a double assumption: it is assumed that in any university community there are serious and capable students whose primary aim is indeed the winning of mastery in a chosen field of knowledge but who at the same time desire to gain some understanding of other fields — some intelligence of their enterprises, their genius, their methods and their achievements; it is further assumed that this non-vocational or avocational propensity is legitimate and laudable. Are the assumptions correct? The latter one involves a question of values and will be dealt with presently. In respect of the former we have to do with what mathematicians call an existence theorem: Do the students described exist? They do. Can the fact be demonstrated — deductively proved? It can not. How, then, may we know it to be true?

The answer is: partly by observation, partly by experience, partly by inference and partly by being candid with ourselves. Who is there among us that is unwilling to admit that he himself now is or at least once was a student of the kind? Where is the university professor to whom such students have not revealed themselves as such in conversation? Who is it that has not learned of their existence through the testimony of others? No doubt some of us not only have known students of the kind, but have tried in a measure to serve them. We may as well be frank. I have myself for some years offered in my subject a course designed in large part for students having no vocational interest in mathematics. I may be permitted to say, for what the testimony may be worth, that the response has been good. The attendance has been composed about equally of students who were not looking forward to a career in mathematics and of students who were. And this leads me to say, in passing, that, if the latter students were asked to explain what value such instruction could have for them, they would probably answer that it served to give them some knowledge *about* a great subject which they could hardly hope to acquire from courses designed solely to give knowledge *of* the subject. Every one knows that it often is of great advantage to treat a subject as an object. One of the chief values of n-dimensional geometry is that it enables us to contemplate ordinary space from the outside, as even those who have but little imagination can contemplate a plane because it does not immerse them. Returning from this digression, permit me to ask: if, without trying to discover the type of student in question, we yet become aware, quite casually, that the type actually exists, is it not legitimate to infer that it is much more numerously represented than is commonly supposed? And if such students occasionally make their presence known even when we do not offer them the kind of instruction to render their wants articulate, is it not reasonable to infer that the provision of such instruction would have the effect of revealing them in much greater numbers?

Indeed it does not seem unreasonable to suppose that a " strong substantial course " of the kind in question, in whatever great subject it were given, would be attended not only by considerable numbers of regular students but in a measure also by officers of instruction in other subjects and even perhaps by other qualified residents of an academic community.   Only the other day one of my mathematical colleagues said to me that he would rejoice in an opportunity to attend such a course in physics.   The dean of a great school of law not long ago expressed the wish that some one might write a book on mathematics in such a way as would enable students like himself to learn something of the innerness of this science, something of its spirit, its range, its ways, achievements and aspiration.   I have known an eminent professor of economics to join a beginner's class in analytical geometry.   Very recently one of the major prophets of philosophy declared it to be his intention to suspend for a season his own special activity in order to devote himself to acquiring some knowledge of modern mathematics.   Similar instances abound and might be cited by any one not only at great length, but in connection with every cardinal division of knowledge.   Their significance is plain.   They are but additional tokens of the fact that the race of catholic-minded men has not been extingiushed by the reigning specialism of the time, but that among students and scholars there are still to be found those whose curiosity and intellectual interests surpass all professional limits and crave instruction more generic in kind, more liberal, if you please, and ampler in its scope, than our vocationalized programs afford.

As to the question of values, I maintain that the desire of such men is entirely legitimate, that it is wholesome and praiseworthy, that it deserves to be stimulated, and that universities ought to meet it, if they can. Indeed, all this seems to me so obvious that I find it a little difficult to treat it seriously as a question. If the matter must be debated, let it be debated on worthy ground. To say, as proponents sometimes say, that, inasmuch as all knowledge turns out sooner or later to be useful, students preparing for a given vocation by specializing in a given field may profitably seek some general acquaintance with other fields *because* such general knowledge will indirectly increase their vocational equipment, is to offer a consideration which, though in itself it is just enough, yet degrades the discussion from its appropriate level, which is that of an ideal humanity, down to the level of mere efficiency and practicianism. No doubt one engaged in minutely studying the topography of a given locality because he intends to reside in it might be plausibly advised to study also the general geography of the globe on the ground that his special topographical knowledge would be thus enhanced, and that, moreover, he might sometime desire to travel. But if we ventured to counsel him so, he might reply: What you say is true. But why do you ply me with such low considerations? Why do you regard me as something crawling on its belly? Don't you know that I ought to acquire a general knowledge of geography, not primarily because it may be useful to me as a resident here or as a possible traveler, but because such knowledge is essential to me in my character as a man? The rebuke, if we were fortunately capable of feeling it, would be well deserved. A man building a bridge is greater than the engineer; a man planting seed is greater than the farmer; a man teaching calculus is greater than the mathematician; a man presiding at a faculty meeting is greater than the dean or the president. We may as well remember that man is superior to any of his occupations. His supreme vocation is not law nor medicine nor theology nor commerce nor war nor journalism nor chemistry nor physics nor mathematics nor literature nor any specific science or art or activity; it is intelligence, and it is this supreme vocation of man as man that gives to universities their supreme obligation. It is unworthy of a university to conceive of man as if he were created to be the servant of utilities, trades, professions and careers: these things are for *him:* not ends but means. It is said that intelligence is good because it prospers us in our trades, industries and professions; it ought to be said that these things are good because and in so far as they prosper intelligence. Even if we do not conceive the office of intelligence to be that of contributing to being in its highest form, which consists in understanding, even if we conceive its function less nobly as that of enabling us to adjust ourselves to our environment, the same conclusion holds. For what is our environment? Is it wholly or mainly a matter of sensible circumstance — sea and land and sky, heat and cold, day and night, seasons, food, raiment, and the like? Far from it. It is rather a matter of spiritual circumstances — ideas, sentiments, doctrines, sciences, institutions, and arts. It is in respect of this ever-changing and ever-developing world of spiritual things, it is in respect of this invisible and intangible environment of life, that universities, whilst aiming to give mastery in this part or that, are at the same time under equal obligation to give to such as can receive it some general orientation in the whole.

And now as to the question of feasibility. Can the thing be done? So far as mathematics is concerned I am confident that it can, and I have a strong lay suspicion that it can be done in all other subjects.

It is my main purpose to show, with some regard to concreteness and detail, that the thing is feasible in mathematics. Before doing so, however, I desire to view the matter a little further in its general aspect and in particular to deal with some of the considerations that tend to deter many scientific specialists from entering upon the enterprise.

One of the considerations, and one, too, that is often but little understood, and so leads to wrong imputations of motive, though it is in a sense distinctly creditable to those who are influenced by it, is the consideration that relates to intricacy and technicality of subject-matter and doctrine. Every specialist knows that the principal developments in his branch of science are too intricate, too technical and too remote from the threshold of the matter to be accessible to laymen, whatever their abilities and attainments in foreign fields. Not only does he know that there is thus but relatively little of his science which laymen can understand but he knows also that the portions which they can not understand are in general precisely those of greatest interest and beauty. And knowing this, he feels, sometimes very strongly, that were he to endeavor by means of a lecture course to give laymen a general acquaintance with his subject, he could not fail to incur the guilt of giving them, not merely an inadequate impression, but an essentially false impression, of the nature, significance and dignity of a great field of knowledge. His hesitance, therefore, is not due, as it is sometimes thought to be, to indifference or to selfishness. Rather is it due to a sense of loyalty to truth, to a sense of veracity, to an unwillingness to mislead or deceive. Of course strange things do sometimes happen, and it is barely conceivable that once in a long time nature may, in a sportive mood, produce a kind of specialist whose subject affects him much as the possession of an apple or a piece of candy affects the boy who goes round the corner in order to have it all himself. But if the type exist, not many men could claim the odd distinction of belonging to it. Specialists are as generous and humane as other men. Their subjects affect them as that same boy is affected when, if he chance to come suddenly upon some strange kind of flower or bird, he at once summons his sister or brother or father or mother or other friend to share in his surprise and joy. There is this difference, however — the specialist must, unfortunately, suffer *his* joy in solitude unless and until he finds a comrade in kind. I admit that the deterrent consideration in question is thoroughly intelligible. I contend that the motive it involves presents an attractive aspect. But I can not think it of sufficient weight to be decisive. It involves, I believe, an erroneous estimate of values, a fallacious view of the ways of truth to men. A few years ago, when making a railway journey through one of the most imposing parts of the Rocky Mountains, I was tempted like many another passenger to procure some photographs of the scenery in order to convey to far-away friends some notion of the wonders of it. So far, however, did the actual scenery surpass the pictures of it, excellent as these were, that I decided not to buy them, feeling it were better to convey no impression at all than to give one so inferior to my own. No doubt the decision might be defended on the ground of its motive. Did it not originate in a certain laudable sense of obligation to truth? Nevertheless, as I am now convinced, the decision was silly. For in accordance with the same principle it is plain that I ought to have wished to have my own impressions erased, seeing that they must have been quite inferior to those of a widely experienced mountaineer as those which the pictures could have given were inferior to mine. Who is so foolish as to argue that no one should learn anything about, say London, unless he means to master all its plans, its architecture and its history in their every phase, feature and detail? Who would contend that because we are permitted to know only so little of what is happening in the European war, we ought to remain in total ignorance of it? Who would say that no one may with propriety seek to learn something about ancient Rome unless he ıs bent on becoming a Gibbon or a Mommsen? It is undoubtedly true that an endeavor to present a body of doctrine or a science to such as can not receive it fully must result in giving a false impression of the truth. But the notion that such an endeavor is therefore wrong is a notion which, if consistently and thoroughly carried out, would put the human mind entirely out of commission. All impressions, all views, all theories, all doctrines, all sciences are false in the sense of being partial, imperfect, incomplete. " Il n'y a plus des problèmes résolus et d'autres qui ne le sont pas, il y a seulement des problèmes *plus ou moins* résolus," said Henri Poincaré. Every one must see that, but for the helpfulness of views which because incomplete are also in a measure false, even the practical conduct of life, not to say the advancement of science, would be impossible. There is no other choice: either we must subsist upon fragments or perish.

Again, many a specialist shrinks from trying to present his subject to laymen because he looks upon such activity as a species of what is called popularization of science, and he believes that such popularization, even in its best sense, closely resembles vulgarization in its worst. He fancies that there is a sharp line bounding off knowledge that is mere knowledge from knowledge that is scientific. In his view science is for specialists and for specialists only. He declines, on something like moral and esthetic grounds, to engage in what he calls playing to the gallery. It might, of course, be said that there is more than one way of playing to the gallery. It could be said that one way consists in acting the rôle of one who imagines that his intellectual interests are so austere and elevated and his thought so profound that a just sense of the awful dignity of his vocation imposes upon him, when in presence of the vulgar multitude, the solemn law of silence. It would be ungenerous, however, if not unfair, to insist upon the justice of such a possible retort. Rather let it be granted, for it is true, that much so-called popularization of science *is* vicious, relieving the ignorant of their modesty without relieving them of their ignorance, equipping them with the vocabulary of knowledge without its content and so fostering not only a vain and empty conceit, but a certain facility of speech that is seemly, impressive and valuable only when, as is too seldom the case, it is accompanied by solid attainments. To say this, however, is not to lay an indictment against that kind of scientific popularization which was so happily illustrated by the very greatest men of antiquity, which was not disdained even by Galileo in the beginnings of modern science nor by

Leonardo da Vinci, and which in our own time has engaged the interest and skill of such men as Clifford and Helmholtz, Haeckel and Huxley, Mach, Ostwald, Enriques and Henri Poincaré.   It is not to arraign that variety of popularization which any one may behold in the constant movement of ideas, once reserved exclusively for graduate students, down into undergraduate curricula and which has, for example, made the doctrine of limits, analytical geometry, projective geometry, and the notions of the derivative and the integral available for presentation to college freshmen or even to high-school pupils.   It is not to condemn that kind of popularization which is so natural a process that it actually goes on in a thousand ways all about us without our deliberate cooperation, without our intention or our consent, and has enriched the common sense and common knowledge of our time with countless precious elements from among the scientific and philosophic discoveries made by other generations of men.

Finally it remains to mention the important type of specialist in whom strongly predominates the predilection for research as distinguished from exposition.   He knows, as every one knows, that through what is called practical applications of science many a scientific discovery is made to serve innumerable human beings who do not understand it and innumerable others who never can.   He may or may not believe in avocational instruction; he may or may not regard intelligence as an ultimate good and an end in itself; he may or may not think that the arts and agencies for the dissemination of knowledge, as distinguished from the discovery and practical applications of truth, are important; he may or may not know that the art and the gifts of the great expositor are as important and as rare as those of the great investigator and less often owe their success to the favor of accident or chance. He may not even have seriously considered these things. He does know his own pre- dilection; and so strong is his inclination towards re- search that for *him* to engage in exposition, especially in popular exposition, in avocational instruction for lay- men, would be to sin against the authority of his vocation. This man, if he have intellectual powers fairly corre- sponding to the seeming authority and urgency of his inner call, belongs to a class whose rights are peculiarly sacred and whose freedom must be guarded in the interest of all mankind. It is not contended that every repre- sentative of a given subject is under obligation to expound it for the avocational interest and enlightenment of lay- men. The contention is that such exposition is so im- portant a service that any university department should contain at least one man who is at once willing and qualified to render it.[2]

I come now to the keeping of my promise. It is to be shown that the service is practicable in the subject of mathematics and how it is so. Let us get clearly in mind the kind of persons for whom the instruction is to be primarily designed. They are to be students of " ma- turity and power " ; they do not intend to become teachers, much less producers, of mathematics; they are probably specializing in other fields; they do not aim at becoming mathematicians; their interest in mathematics is not vocational, it is avocational; it is the interest of those whose curiosity transcends the limits of any specific profession or any specific form or field of activity; each of them knows that, whatever his own field may be, it is

[2] Concerning the duty and failure of scientific men to enlighten the public see the chapter on Non-Euclidean Geometry in Keyser's *Mathematical Philosophy,* Dutton and Company.

penetrated, overarched, compassed about by an infinitely vaster world of human interests and human achievements; they feel its immense presence, the poignant challenge of it all; as specialists they will win mastery over a little part, but they have heard the call to intelligence and are seeking orientation in the whole; this they know is a thing of mind; they are aware that the essential environment of a scholar's life is a spiritual environment — the invisible and intangible world of ideas, doctrines, institutions, sciences and arts; they know or they suspect that one of the great components of that world is mathematics; and so, not as candidates for a profession or a degree, but in their higher capacity as men and women, they desire to learn something of this science viewed as a human enterprise, as a body of human achievements; and they are willing to pay the price; they are not seeking entertainment, they are prepared to work — to listen, to read and to think.

And now we must ask: What measure of mathematical training is to be required of them as a preparation? In view of what has just been said it is evident that such training is not to be the whole of their equipment nor even the principal part of it, but it is an indispensable part. And the question is: How much mathematical knowledge and mathematical discipline is to be demanded? I have no desire to minimize my present task. I, therefore, propose that only so much mathematical preparation shall be demanded as can be gained in a year of collegiate study. Most of them will, of course, have had more; but I propose as a hypothesis that the amount named be regarded as an adequate minimum. But it does not include the differential and integral calculus. And is it not preposterous to talk of offering graduate instruction in mathematics to students who have not had a first course in the calculus? I am far from thinking so. A little reflection will suffice to show that in the case of such students as I have described it is very far from preposterous. In my opinion the absurdity would rather lie in demanding the calculus of them. No one is so foolish as to contend that a first course in the calculus is a *sufficient* preparation for undertaking the pursuit of graduate mathematical study. But to suppose it necessary is just as foolish as to suppose it sufficient. There was a time when it *was* necessary, and the belief that it is necessary now owes its persistence and currency to the inertia then acquired. Formerly it was necessary, because formerly all advanced courses, at least all initial courses of the kind, were either prolongations of the calculus, like differential equations, for example, or else courses in which the calculus played an essential instrumental rôle as in rational mechanics, or the usual introductions to function theory or to higher geometry or algebra. But, as every mathematician knows, that time has passed. It is true that courses for which a preliminary training in the calculus is essential still constitute and will continue to constitute the major part of the graduate offer of any department of mathematics. And quite apart from that consideration, it seems wise, in the case of intending graduate students who purpose to specialize in mathematics, to enforce the usual calculus requirement as affording some slight protection against immaturity and the lack of seriousness. But every mathematician knows that it is now practicable to provide a large and diversified body of genuinely graduate mathematical instruction for which the calculus is strictly not prerequisite.

Fortunately it is just the material that is thus available which is in itself best suited for the avocational instruction we are contemplating. As the calculus is not to be presupposed it goes without saying that this subject must find a place in the scheme. For evidently an advanced mathematical course devised and conducted in the interest of general intelligence can not be silent respecting " the most powerful weapon of thought yet devised by the wit of man." Technique is not sought and can not be given. The subject is not to be presented as to undergraduates. For the most part these gain facility with but little comprehension. It is to be presented to mature and capable students who seek, not facility, but understanding. Their desire is to acquire a general conception of the nature of the calculus and of its place in science and the history of thought — such a conception as will at least enable them as educated men to mention the subject without a feeling of sham or to hear it mentioned without a feeling of shame. A few well-considered lectures should suffice. At all events it would not require many to show the historical background of the calculus, to explain the nascence and nature of the scientific exigencies that gave it birth, to make clear the concepts of derivative and integral as the two central notions of its two great branches, and to present a few simple applications of these notions to intelligible problems of typical significance. Even the idea of a differential equation could be quickly reached, the nature of a solution explained, and simple examples given of physical and geometric interpretations. As to the range and power of the calculus, a sense and insight can be given, in some measure of course by a reference to its literature, but much more effectively by a few problems carefully selected from various fields of science and skillfully explained with a view to showing wherein the methods of the calculus are demanded and how they serve. Is not all this elementary and undergraduate? In point of nomenclature, yes. It is not necessary, however, to let words deceive us. We teach whole numbers to young children, but even Weierstrass was not aware of the logico-mathematical deeps that underlie cardinal arithmetic.

The calculus, however, is hardly the topic with which the course would naturally begin. A principal aim of the course should be to show what mathematics, in its inner nature, is — to lay bare its distinctive character. Its distinctive character, its structural nature, is that of a " hypothetico-deductive " system. Probably, therefore, it would be well to begin with an exposition of the nature and function of postulate systems and of the great rôle such systems have always played in the science, especially in the illustrious period of Greek mathematics and even more consciously and elaborately in our own time. It is plain that such an exposition can be made to yield fundamental insight into many matters of interest and importance not only in mathematics, but in logic, in psychology, in philosophy, and in the methodology of natural science and general thought. The material is almost superabundant, so numerous are the postulate systems that have been devised as foundations for many different branches of geometry, algebra, analysis, *Mengenlehre* and logic. A general survey of these, were it desirable to pass them all in review, would not be sufficient. It will be necessary to select a few systems of typical importance for minute examination with reference to such capital points as convenience, simplicity, adequacy, independence, compatibility and categoricalness. The necessity and presence of undefined terms in any and all systems will afford a suitable opportunity to deal with the highly important, much neglected and little understood subject of definition, its nature, varieties and function, in light of the recent literature, especially the suggestive handling of the matter by Enriques in his "Problems of Science." A given system once thus examined, the easy deduction of a few theorems will suffice to show the possibility and the process of erecting upon it a perfectly determinate and often imposing superstructure. And so will arise clearly the just conception of a mathematical doctrine as a body of thought composed of a few undefined together with many defined ideas and a few primitive or postulated propositions with many demonstrated ones, all concatenated and welded into a form independent of will and temporal vicissitudes. Revelation of the charm of the science will have been begun. A new revelation will result when next the possibility is shown of so interchanging undefined with defined ideas and postulates with demonstrated propositions that, despite such interchange of basal with superstructural elements, the doctrine as an autonomous whole will remain absolutely unchanged. But this is not all nor nearly all. It is only the beginning of what may be made a veritable apocalypse. Of great interest to any intellectual man or woman, of very great interest to students of logic, psychology, or philosophy, should be the light which it will be possible in this connection to throw upon the economic rôle of logic and upon the constitution of mind or the world of thought. I refer especially to the recently discovered fact that in interpreting a system of postulates we are not restricted to a single possibility, but that, on the contrary, such a system admits in general of a literally endless variety of interpretations; which means, for such is the makeup of our *Gedankenwelt*, that an infinitude of doctrines, widely different in respect of their psychological character and interest, have nevertheless a common form, being isomorphic, as we say, logically one, though spiritually many, reposing on a single base.

And how foolish the instructor would be not to avail himself of the opportunity of showing, too, in the same connection, how various mathematical doctrines that differ not only psychologically, but logically also, are yet such that, by virtue of a partial agreement in their bases, they intersect one another, owning part of their content jointly, whilst being, in respect of the rest, mutually exclusive and incompatible. If, for example, it be some Euclidean system that he has been expounding, he will be able readily to show upon how seemingly slight changes of bases there arise now this or that variety of non-Euclidean geometry, now a projective or an inversion geometry or some species or form of higher dimensionality. I need not say that analogous phenomena will in like manner present themselves in other mathematical fields. And it is of course obvious that as various doctrines are thus made to pass along in deliberate panorama it will be feasible to point out some of their salient and distinctive features, to indicate their historic settings, and to cite the more accessible portions of their respective literatures. Naturally in this connection and in the atmosphere of such a course the question will arise as to why it is that, or wherein, the hypothetico-deductive method fails of universal applicability. So there will be opportunity to teach the great lesson that this method is not rudimentary, but is an ideal, the ideal of intellect and science; to teach that mathematics is but the name of its occasional realization; and that, though the ideal is, relatively speaking, but seldom attained, yet its lure is universal, manifesting itself in the most widely differing domains, in the physical and mechanical assumptions of Newton, in the ethical postulates of Spinoza, in our federal constitution, even in the ten commandments, in every field where men have sought a body of principles to serve them as a basis of doctrine, conduct or achievement. And if it shall thus appear that mathematics is very high-placed as being, in respect of its method and its form, the ideal and the lure of thought in general, the fault must be imputed, not to the instructor, but to the nature of things.

In all this study of the postulational method the impression will be gained that the science of mathematics consists of a large and increasing number of more or less independent, somewhat closely related and often interpenetrating branches, constituting, not a jungle, but rather an immense, diversified, beautifully ordered forest; and that impression is just. At the same time another impression will be gained, namely, that the various branches rest, each of them, upon a foundation of its own. This impression will have to be corrected. It will have to be shown that the branch-foundations are not really fundamental in the science but are literally and genuinely component parts of the superstructure. It will have to be shown that mathematics as a whole, as a single unitary body of doctrine, rests upon a basis of primitive ideas and primitive propositions that lie far below the so-called branch-foundations and, in supporting the whole, support these as parts. The course will, therefore, turn to the task of acquainting its students with those strictly fundamental researches which we associate with such names as C. S. Peirce, Schroeder, Peano, Frege, Russell, Whitehead and others, and which have resulted in building underneath the traditional science a logico-mathematical sub-structure that is, philosophically, the most important of modern mathematical developments.

It must not be supposed, however, that the instruction must needs be, nor that it should preferably be, confined to questions of postulate and foundation, and I will devote the remainder of the time at my disposal to indicating briefly how, as it seems to me, a large or even a major part of the course may concern itself with matters more traditional and more concrete.

Any one can see that there is an abundance of available material. There is, for example, the history and significance of the great concept of function, a concept which mathematics has but slowly extracted and gradually refined from out the common content and experience of all minds and which on that account can be not only defined precisely and intelligibly to such laymen as are here concerned, but can also be clarified in many of its forms by means of manifold examples drawn from elementary mathematics, from the elements of other sciences, and from the most familiar phenomena of the work-a-day world.

Another available topic is the nature and rôle of the sovereign notion of limit. This, too, as every mathematician knows, admits of countless illustration and application within the radius of mathematical knowledge here presupposed. In this connection the structure and importance of what Sylvester called " the Grand Continuum," which so many scientific and other folk talk about unintelligently, will offer itself for explanation. And if the class fortunately contain students of philosophic mind, they will be edified and a little astonished perhaps when they are led to see that the method and the concept of limits are but mathematicized forms of a process and notion familiar in all domains of spiritual activity and known as idealization. Not improbably some of the students will be sufficiently enterprising to trace the mentioned similitude in some of its manifestations in natural science, in psychology, in philosophy, in jurisprudence, in literature and in art.

I have not mentioned the modern doctrine variously known as *Mengenlehre, Mannigfaltigkeitslehre*, the theory of point-sets, assemblages, manifolds, or aggregates: a live and growing doctrine in which expert and layman are about equally interested and which, like a subtle and illuminating ether, is more and more pervading mathematics in all its branches. For the avocational instruction of lay students of " maturity and power " how rich a body of material is here, with all its fascinating distinctions of discrete and continuous, finite and infinite, denumerable and non-denumerable, orderless, ordered, and with its teeming host of near-lying propositions, so interesting, so illuminating, often so amazing.

Finally, but far from exhausting the list, it remains to mention the great subjects of invariants and groups. Both of them admit of definition perfectly intelligible to disciplined laymen; both admit of endless elementary illustration, of having their mutual relations simply exemplified, of being shown in historic perspective, and of being strikingly connected, especially the notion of invariance, with the dominant enterprise of man: his ceaseless quest for the changeless amid the turmoil and transformation of the cosmic flux.[3]

[3] My *Mathematical Philosophy* (Dutton & Co.) was written to afford educated laymen the kind of instruction advocated ten years earlier in the foregoing address.

THE SOURCE AND FUNCTIONS OF A UNIVERSITY [1]

In returning hither from near and far to join in celebrating the seventy-fifth anniversary of the founding of their academic birthplace and home, the alumni, the sons and daughters of this institution, have not come to congratulate an eld-worn mother upon the continuance of her years beyond the Psalmist's allotment of three score and ten nor to comfort her in the sorrows of age. Their assembling is due to other sentiments and owns another mood. They have come as beneficiaries in order to pay, for themselves and for the many absent ones whom they have the honor to represent, a tribute of gratitude, loyalty and love to a noble benefactress who, notwithstanding her wisdom and fame, yet is literally in the early morning of her life. For it is not written, or ordained in the scheme of things, that, in respect of years, the life of a university shall be as a tale that is told or a watch in the night. It is indeed a living demonstration of the greatness of man, bearing witness to his superiority even over death, that men and women, though they themselves must die, yet may, whilst they live, create ideals and institutions that survive. A college or a university may indeed have been as a benignant mother to a thousand academic generations and yet be younger than her youngest child. Unlike man the individual, a university is, like man the

[1] An address delivered June 3, 1914, at the celebration of the seventy-fifth anniversary of the founding of the University of Missouri. Printed in *The Columbia University Quarterly*, March, 1915.

race, immortal. The age of three score and fifteen in
the life of an immortal institution is a mere beginning.
In emphasizing this consideration it is not my intention
to suggest or imply that the services rendered by the
University of Missouri have necessarily been, because
of her youth, meagre or ineffectual or immature. On
the contrary I maintain that her services to the people of
this state have been beyond computation and that already
her spiritual achievements constitute the chief glory of a
great commonwealth. Is it the alumni only who owe her
grateful allegiance? Is the beneficence of an institution of learning exclusively or even mainly confined to
the relatively few who dwell for a season in her immediate presence, who touch the hem of her garment, come
into personal contact with her scholars and teachers and
receive her degrees? Far from it. Far from being the
sole or the principal beneficiaries of a university, the
alumni are simply among the more potent instrumentalities for extending her ministrations to ever wider and
wider circles. The sun, we say, is far off yonder in the
heavens. But strictly speaking the sun really *is* wherever
he shines. Where is the University of Missouri? At
Columbia, we say, and the speech is convenient. But
it is juster to say that, owing to the pervasiveness of her
light and inspiration, the University of Missouri in a
measure now is, and in larger and larger measure will
come to be, in every home and school, in every factory
and field, in every mine and shop, in every council chamber, in every office of charity, or medicine, or law, in all
the places near or remote where within the borders of
this beautiful state children play and men and women
think and love, suffer and hope, aspire and toil. Nay,
by the researches and publications of her scholars and
by the migrations of those she has inspired and dis-

ciplined, the University of Missouri to-day lives and moves abroad, mingling her presence with that of kindred agencies, not only in every state of the union but in many other quarters of the civilized world.

It is not my purpose to review the history of her aspirations and struggles nor to relate the thrilling story of her triumphs. I conceive that the central motive of our assembling here is not so much to praise the University for what she has already accomplished as to renew our devotion to her high emprize, to congratulate her upon her solid attainments, to rejoice in her divine discontent and spirit of progressiveness, to deepen and enlarge our conception of her mission and destiny, and especially to remind ourselves of the principles, the faith and, above all, the ideals to which she owes her birth, her continuity, her responsibilities, and her power.

What is a university? How shall we conceive that marvelous thing which, though having a local habitation and a name and seeming to dwell in houses made by human hands, yet contrives to be omnipresent; pervading the abodes of men everywhere throughout a state, a nation or a world, like a divine ether; subtly, gently, unceasingly, increasingly ministering to their hearts and minds healing counsels and the mysterious grace of light and understanding? What is it? Is it something, an agency or an influence, that can be defined? We know that it is not. We know that the really great things of the world, the things that live and grow and shine, the things that give to life its interests and its worth, one and all elude formulation. Yet it is just these things, beauty and love, poetry and thought, religion and truth and mind, it is precisely these great indefinables of life that we may learn, through experience and discipline, to know best of all. And so it is with what we mean or ought to mean by a university. What a university is no one can define, but all may in a measure come to know. By pondering its principles, by contemplating its ideals, by examining its aims, activities and fruits, above all by sharing in its spirit and aspirations, we may at length win a conception of it that will fill our minds with light and our hearts with devotion.

Where such a conception reigns a university will flourish. But there is no conception more difficult for a people to acquire. It is not a spontaneous growth, springing up like a weed, but requires careful planting and cultivation. Such is the husbandry to which a university must perpetually devote itself as the essential precondition to the prosperous exercise and advancement of all its other functions, and the husbandry is not easy. Especially in our American communities where universities must appeal for support to the intelligence of a democratic people, there is no service more important or more difficult to render than that which consists in teaching us to know what a university really is and what it signifies alike for developing the material resources of the world and for the spiritualizing of man. And thus there devolves upon a university, especially in the beginning of its career, the necessity of performing a kind of miracle: without adequate support, either material or moral, it must yet find strength to teach us to give it both. The lesson is one that takes long and long to teach because it is one that takes long and long to learn.

It is a great mistake to imagine that a university is an essentially modern thing. In spirit, in idea and essence, it is modern only in the sense in which forces and ideals that are eternal are always modern, as they are always ancient. We should not forget that even the name University — so suggestive of the infinite world which it is the aim of these institutions to subjugate to the understanding and uses of man — even the name, in its modern scholastic sense, has had a history of more than a thousand years. But we know that the institution itself, the thing that bears the name, owns an antiquity far more remote. A few years ago, standing upon the Acropolis of Athens, gazing pensively about upon the hallowed scene where culminated the genius of the ancient world, a friend, pointing towards the spot near by where for fifty years Plato taught in the grove of Academe, said to me, yonder, yonder is the holy ground where was made the first attempt to organize higher education in the western world. The remark, which was just enough, was indeed impressive. It is easy, however, to misunderstand its significance and to exaggerate its importance. So many of the most precious elements of our civilization trace their lineage back to the creative activity of ancient Greece that we are naturally tempted to imagine we may find there also the source and origin of those aims, activities and ideals which constitute what we today call a university. Such imagining, however, is vain. The originals, the first organizations, we may possibly find there at a definite time and place, but not the origin, not the source, not the nascence of the birth-giving and life-sustaining power. For this must account, not only for the universities of our time, but for the great school of Plato as well. What, then, and where is the secret spring?

Shall we seek it in a sense of need? Necessity is indeed a keen spur to invention and is the mother of many things. But necessity is not the mother of universities. The beasts flourish and propagate their kind without the help of institutions of learning, and without such help a similar existence is possible to men. Universities are not essential to life nor to animal prosperity. They are not creatures, they are creators, of need. We do indeed nowadays hear much of the services they render, and it is right that we should, for they minister constantly and everywhere to countless forms of need. But the needs they supply are in the main needs that they have first produced, multiplied desires and aspirations, new propensions of mind awakened to new life, lifted by education to higher levels and ampler possibilities of being. No, the origin, the source we are seeking, the principle of explanation, is no human contrivance nor institution nor sense of need. It is that sovereign urgency, at once so strange and so familiar, that drives us to seek it; it is the lure of wisdom and understanding, of beauty and light; a certain divine energy in the world, at once a cosmic force and a human faculty, constituting man divine in constituting him a seeker of truth and a lover of harmony and illumination.

Has it an epoch and a name? It has both. In accordance with the modern doctrine of evolution the greatest events upon our planet occurred long before the beginnings of recorded history. For according to that doctrine there must have come a time, long, long ago, when in what was a world of matter there began to be mind, in what was a world of motion there began to be emotion, and the blind dominion of force was invaded by personality. Among all those marvels of prehistoric history, the supreme event was that one but for which this world had been a world devoid of mystery and devoid of truth — I mean the advent of Wonder. With the advent of wonder came the sense of mystery, the lure of truth, the sheen of ideality, the dream of the perfect and, with these, the potence and promise of research and creativeness with all their endless progeny of knowledge and wisdom and science and art and philosophy and religion. These things, children of the spirit, offspring of wonder, these things are the interests which it is the divine prerogative of universities to serve, and the universities ultimately derive their own existence, their sustenance and their power from the same mother that gives their charges birth. A genuine university is thus the offspring and the appointed agent of the spirit of inquiry; it is the offspring, expression and servant of that imperious curiosity which in a measure impels all men and women, but with an urgency like destiny literally *drives* men and women of genius, to seek to know and to teach to their fellows whatsoever is worthy in all that has been discovered or thought, spoken or done in the world, and at the same time seeks to extend the empire of understanding endlessly in all directions throughout the infinite domain of the yet uncharted and unknown. That high commission is at once a university's charter of freedom and the definition of her functions and her obligations. These are, on the one hand, to teach — to teach with no restrictions save those prescribed by decency and candor — and, on the other hand, to foster and prosecute research — research in any and all subjects or fields to which the leading or the stress of curiosity may draw or impel. In so far as the great commonwealth of Missouri makes ample provision for the exercise of these functions and for the discharge of these obligations, to that extent she may be said to cooperate with the divine energy of the world in the maintenance of a genuine university.

RESEARCH IN AMERICAN UNIVERSITIES [1]

THE present writer has been asked to deal briefly with the question of research in American universities. The subject is an immense one, and the following discussion makes no pretense of being exhaustive. It aims merely to present the problem again, to emphasize again its importance, and to point out once more some of its harder conditions and some of the principles and distinctions involved in any serious attempt at its solution.

The problem may not be easy to appreciate, but it is at all events easy to state. It is the problem of securing in our universities suitable provision for the work of research or investigation and productivity. For a generation the great majority of the ablest men in our universities have regarded that problem as the most urgent and important educational problem confronting these institutions and the American people. Meanwhile, something has been done towards a solution. But none of the universities has secured adequate provision, and the majority of them but little or none at all. In the abstract, the problem is simple and the solution is easy: given a body of able and enthusiastic men, provide them with proper facilities, afford them opportunity to devote their powers continuously to the prosecution of research, and the thing is done. But in the concrete it is exceedingly difficult, being frightfully complicated with our whole institutional history and life, in particular with our educational traditions and tendencies, with the prevailing plan

[1] Printed in *The Bookman*, May, 1906.

of university organisation, and especially with the characteristic temper, ideals and ambitions of the American
people.

Somebody besides our foreign friends and critics ought
to tell the truth about American education and American
universities. Our people have never ceased to believe in
education. Our belief has not always been intelligent.
We have been prone to ascribe to education efficacies and
potencies that do not belong to any human agency or institution. But our faith in it, though not always critical
or enlightened, has been deep, implicit and abiding; and
we have diligently pursued it, generally as a means no
doubt, but sometimes as an end, and occasionally as a
thing in itself more precious than power and gold. In all
this we have been, quite unconsciously and contrary to
all appearances, very humble. We have been content to
educate ourselves with knowledge discovered by others
and to nourish ourselves with doctrines and truths produced only by the spiritual activity of other lands. We
may have been vain but we have not been proud. Besides
a marvelous practical sense we have had, in degree quite
unsurpassed, two of the elements of genius, — intellectual
energy and intellectual audacity; and by means of these
we have created a material civilisation so obtrusive, so
elaborate and so efficient as to amaze the world. But
now at length there begin to appear the indicia of change,
of change for the better. A new day has dawned. The
sun is not yet risen high, but it is rising. We have begun
to suspect that genuine civilisation is essentially an affair
of the spirit, that it can not be borrowed nor imported nor
improvised nor appropriated from without, but that it
is a growth from within, an efflorescence of mind and
soul, and that its highest tokens are not soldiers but
savants, not the purchasers and admirers of art but

artists, not mere retailers of knowledge nor teachers of the familiar and the known, but discoverers of the unknown, not mere inventors but men of science. And so we have begun to feel our way towards the establishment of true universities, that is to say of institutional centres for the activity of the human spirit, and of organs, the most potent yet invented by human society, for giving effect to the noblest instinct of man, " the civilisation-producing instinct of truth for truth's sake."

Just here we encounter a great danger. For a generation our progress in the matter has been so swift that both the universities themselves and the educated public opinion upon which in our democratic society their support and advancement ultimately depend, are in danger of greatly overestimating it, and that would be a misfortune. Absolutely the progress has indeed been great, but relatively and judged by the very highest standards, it has not. It is not first nor mainly a question of achievements, of things done. It is a question of ideals, of standards and aspirations. A clear concept of a great university unconsciously serving the highest interests of man by absolute devotion to Truth for its own sake and without extraneous motive, end or aim, does not yet exist in the mind of the American public and is not yet incarnate in any of its institutions. Our universities are young, strong and robust. They are full of potence and promise. But they have not yet impressed their own imperfect ideals upon the people; they have not yet given forth the light necessary for their own proper beholding and appreciation. Their perfections and their imperfections alike, remain obscure. The old colleges about which as about nuclei some of our universities have been formed have done much to leaven and temper the American mind and to subdue it to the influences of beauty and truth.

Corresponding services have not yet been rendered by our universities as such. No one can doubt that they are destined to assume in future the permanent leadership, and to exercise a controlling formative influence, in all that goes to deepen thought and to exalt and refine standards, character, and taste. At present, however, they are themselves in the formative and impressionable stage, resembling improvisations in some respects; and to understand them, to see clearly both what they are and what they are not, it is necessary to regard them as being at the present time less the producers than the products of our civilisation.

So regarded, they are seen to embody and to reflect alike the merits and the defects of their progenitor. Like the latter they are unsurpassed in boldness, in energy and in enthusiasm, and their genius has been mainly directed to material and outer ends. Their first and chief concern has been with the physical and exterior, with buildings and grounds and instruments and laboratories, and while their material equipment is still far from adequate, it has already evoked astonished and admiring commentary from visiting scholars of European seats of learning. Like the civilisation whence they have sprung, our universities are intensely modern and up-to-date, and they are intensely democratic in everything but management; they set great store by organisation, exalt the function of administration, and tend to be regarded, to regard themselves, and in fact to be, as vast and complicate machines or industrial plants naturally demanding the control of centralised authority. They have but little sentiment; they are almost devoid of sacred and hallowing traditions, of great and illustrious recollections; there is in and about them nothing or but little of " the shadow and the hush of a haunted past." They have no an-

tiquity. In them the utilitarian spirit, having learned the
lingo of service, contrives to receive an ample share of
honour, and the Genius of Industry that has transformed
our land into an abode of wealth and for generations
assigned an attainable upper limit to a people's aspiration,
shapes educational policy, holds and wields the balance
of power. The classic distinctions of good, better and best
in subjects and motives of study receive but slight regard.
The traditional hierarchy of educational values and the
ascending scale of spiritual worths have fallen into dis-
repute. All things have been leveled up or leveled down
to a common level; so that the workshop and the laboratory, schools of engineering, of agriculture and of the
classics, the library, the model dairy and departments of
architecture and music, exist side by side. In at least
one institution, so it is reported, the professor of poetry
rubs shoulders with the professor of poultry. No wonder
that a distinguished critic has said that some of our
biggest universities appear as hardly more than episodes
in the wondrous maelstrom of our industrial life.

Thus it appears that the American university, child of
a predominantly material and industrial civilisation half-
blindly aspiring to higher things, strikingly resembles its
parent. Begotten in the hope that it would be as a
saviour and rescue us from our national idols and respectable sins, it straightway became their most enlightened
servant and lent them the sanction and the support of its
honoured name. It is by no means contended that this
fact is the whole truth. Our universities are not entirely
devoted to the service of industry; they are not wholly
committed to teaching youth the known from utilitarian
motives and for immediate and practical ends; they are
not exclusively concerned with the *applications* of science; out of general devotion to the Useful, something is

saved for the True; science is not always regarded as a
commodity; the judgment of the great Jacobi is sometimes recognised as just: " The unique end of science is
the honour of the human spirit." And it is a pleasure to
be able to proclaim the fact that in a few of our universities something like a home has been provided for the
spirit of research and that by its activity there, American
genius has had a share in extending the empire of light,
in enlarging the domain of the known, in astronomy, in
physics, in mathematics, in the science of mind, in biology, in criticism, in economics, in letters, in almost all of
the great fields where the instinct of truth for the sake
of truth contends against the dark. In this clear evidence
of our growing freedom and exaltation, let us rejoice; but
let us be candid also. Let us admit that we have only
begun the higher service of the soul; let us confess in
becoming humility that, in comparison with our wealth,
our numbers, our energies and our talents, in comparison,
too, with the intellectual achievements of some other
peoples and other lands, the service we have rendered to
Science and Art and Truth is meagre.

Why such emptiness, such poverty, such meagreness in
the fruits of the highest activity? The immediate cause
is easy to find. It is not incompetence nor lack of genius
in our university faculties. These are not inferior to the
best in the world. It is not mainly due, as is often said,
to inadequacy of material compensation, though one of
the greatest of living physicists, Professor J. J. Thomson,
has told us truly that American men of science receive
less remuneration than their colleagues in any other part
of the world. The cause in question is simple: *lack of
opportunity*. The difficulty is near at hand. It inheres
in the composition and organisation of our universities.
Most of these are built about and upon, and largely con-

sist of, immense undergraduate schools thronged by young men mainly bent upon practical aims and neither qualified nor intending to qualify for the work of investigation. The interests of these schools are naturally the paramount concern. The great and growing burdens of administration tend to distribute themselves among the professors. These have, besides, to give the most and the best of their energies to elementary teaching, to teaching, that is, which does not pertain to a university proper but to gymnasia and lycées — a worthy, important, necessary kind of work, but a kind that drains off the energy in non-productive channels and tends to form and harden the minds of those engaged in it about a small group of simpler ideas. What is left, what can be left, of spirit, of energy, of opportunity, for the arduous work of research? One man attempting the enterprise of three: administration, elementary teaching, discovery and creative work. Who can suitably characterise the absurdity? Who can compute the wickedness of the waste in the impossible attempt to effect daily the demanded transition from mood to mood? A mind, by prolonged effort, at length immersed in the depths of a profound and difficult investigation — how poignant the pain of interruption, the rending of continuity, the rude disturbance of poise and concentration. How easy to fail of due respect for, because it is so easy not to understand, the creative mood, oblivious to the outer world, the brooding " maternity of mind," more delicate than fabric of gossamer, of infinite subtlety, of infinite sensitiveness, of woven psychic structure finer than ether threads; and how easy to forget that a sudden alien call may disturb and jar and even destroy the structure.

Little excuse, then, have we to wonder at the recent words of Professor Bjerknes, of the chair of mechanics

and mathematical physics in the University of Stockholm,
and non-resident lecturer in mathematical physics in
Columbia University, who, in his farewell address to his
American colleagues, assembled to do him honour, spoke
substantially as follows:

" I have been much impressed with the material equipment of your
universities, with your splendid buildings, with the fine instruments you
have placed in them, and with the enthusiasm of the men I have found
at work there. But I hope you will pardon me, gentlemen, for saying,
as I must say, that, when I found you attempting serious investigation
with the remnants of energy left after your excessive teaching and
administrative work, I could not help thinking you did not appreciate
the fact that the finest instruments in those buildings are your brains.
I heard one of you counsel his colleagues to care for the astronomical
instruments lest these become strained and cease to give true results.
Allow me to substitute brain for telescope, and to exhort you to care for
your brains. I have been astonished to find that some of you, in
addition to much executive work, teach from ten to fifteen and even
more hours per week. I myself teach two hours per week, and I can
assure you that, if I had been required to do so much of it as you do,
you never would have invited me to lecture here in a difficult branch
of science. That, gentlemen, is the most important message I can leave
with you."

Such, then, is the situation. No need that we should
behold it in picture drawn by foreign hand. We need
no copy. The original lies before us in all its proportions.
The challenge addresses itself at once to our pride and
to our practical sense. Of all peoples, we, it would seem,
should feel the challenge most keenly, for the problem is
a problem in freedom. It demands the emancipation of
American genius; it demands provision of free and ample
opportunity for the highest activity of our highest talent.

Hope of solution lies in division of labour. Our universities and the people they represent must reduce their
exactions. For three men's work, three must be provided.
There must be men to administer and men to teach and
men to investigate. Three varieties of service, entirely

compatible in kind, entirely incompatible as co-ordinate vocations combined in one. Any one of them may be as an avocation to another of the three, but only so of choice and not by compulsion. No invidious comparisons are implied. The distinctions are not of greater and less; they are matters of economy in the domain of mind. The great administrator is not a clerk nor an amanuensis; he is a man of constructive genius, a creator. The great teacher is not a pedagogue; he is a source of inspiration and of aspiration, producing children of the spirit by " the urge and ardor " of a deep and rich and enlightened personality; he was in the mind of Goethe when he said of Winckelmann that " from him you learned nothing, but you became something." And the great investigator is not a mere collector and recorder of facts; he is a discoverer, a discloser, of the harmonies and the invariance hid beneath the surface of seeming disorder and of ceaseless change. The three great powers are compatible, and are usually found united in a single gigantic personality, just as the ordinary administrator and ordinary teacher and ordinary investigator compose one unit of mediocrity.

It is perfectly evident that the total service demanded of the universities will not diminish. On the contrary, it will continue as now to increase in response to growing need. The case, then, is clear: the number of servants must be increased, the number of those who are to do the work must be greatly multiplied. And thus the problem becomes a financial one. But a university is not a money-making institution. Its function is to convert the physical into the spiritual, to transform the things of matter into the things of mind. It has, however, a physical body, without which it may not dwell among men; and, for the support of it, it depends and must depend, whether through legislative appropriation or the benefaction of individuals, ultimately upon the people. These now possess the means in ample measure, and the promptings of generosity are in the hearts of many wealthy and sagacious men.

And so the problem revolves upon itself and once more turns full upon us its theoretic aspect. Its solution awaits public appreciation of its significance and its terms. It is above all else a question of enlightenment. Just here, if I am not mistaken, is the measureless opportunity of the university president. Beyond all others, he is spokesman and representative before the people of their highest spiritual interests. Their ideals and aspirations will scarcely surpass his own. The problem must be conceived boldly in truth and presented in its larger aspects. It must be seen and be felt to be the supreme problem of our civilisation. As a people we have yet to learn the lesson deeply that research, the competent application in any field whatever of human interest of any effective method whatever for the discovery of truth and enlarging the bounds of knowledge, is the highest form of human activity. We have yet to learn that a nation, a state, a university without investigators, is a community without men of profoundest conviction. For this can not be gained by conning books; it can not be inherited; it is not merely a pious hope or a pleasing superstition. It is not an obsession.

As Helmholz has said, a teacher " who desires to give his hearers a perfect conviction of the truth of his principles must, first of all, know from his own experience how conviction is acquired and how not. He must have known how to acquire conviction where no predecessor had been before him — that is, he must have worked at the confines of knowledge and have conquered new regions." We have yet to learn that the value of a university professor can not be estimated by counting the hours he stands before his classes. We have yet to learn to prefer standards of quality to units of quantity. We have yet to learn that the spirit of pure research, the highest productive genius, has no direct concern whatever with the useful; that, while it does without intention create an atmosphere in which utilities most greatly flourish, it is itself concerned solely with the true; we have yet to learn that " the action of faculty is imperious and always excludes the reflection *why* it acts." When these and kindred lessons shall have been taken to heart, our emancipation, now well begun, will advance towards completion; the American university will come to its own; and our present civilisation will speedily pass to the rank of the highest and best.

PRINCIPIA MATHEMATICA [1]

MATHEMATICIANS, many philosophers, logicians and physicists, and a large number of other people are aware of the fact that mathematical activity, like the activity in numerous other fields of study and research, has been in large part for a century distinctively and increasingly critical. Every one has heard of a critical movement in mathematics and of certain mathematicians distinguished for their insistence upon precision and logical cogency. Under the influence of the critical spirit of the time mathematicians, having inherited the traditional belief that the human mind can know some propositions to be true, convinced that mathematics may not contain any false propositions, and nevertheless finding that numerous so-called mathematical propositions were certainly not true, began to re-examine the existing body of what was called mathematics with a view to purging it of the false and of thus putting an end to what, rightly viewed, was a kind of scientific scandal. Their aim was truth, not the whole truth, but nothing but truth. And the aim was consistent with the traditional faith which they inherited. They believed that there were such things as self-evident propositions, known as axioms. They believed that the traditional logic, come down from Aristotle, was an absolutely perfect machinery for ascertaining what was involved in the axioms. At this stage, therefore, they believed that, in order that a given branch of

[1] An account of Messrs. Whitehead and Russell's great work bearing this title. Printed in *Science,* vol. xxv.

mathematics should contain truth and nothing but truth, it was sufficient to find the appropriate axioms and then, by the engine of deductive logic, to explicate their meaning or content. To be sure, one might have trouble to " find " the axioms and in the matter of explication one might be an imperfect engineer; but by trying hard enough all difficulties could be surmounted for the axioms existed and the engine was perfect. But mathematicians were destined not to remain long in this comfortable position. The critical demon is a restless and relentless demon; and, having brought them thus far, it soon drove them far beyond. It was discovered that an axiom of a given set could be replaced by its contradictory and that the consequences of the new set stood all the experiential tests of truth just as well as did the consequences of the old set, that is, perfectly. Thus belief in the self-evidence of axioms received a fatal blow. For why regard a proposition self-evident when its contradictory would work just as well? But if we do not know that our axioms are true, what about their consequences? Logic gives us these, but as to their being true or false, it is indifferent and silent.

Thus mathematics has acquired a certain modesty. The critical mathematician has abandoned the search for truth. He no longer flatters himself that his propositions are or can be known to him or to any other human being to be true; and he contents himself with aiming at the correct, or the consistent. The distinction is not annulled nor even blurred by the reflection that consistency contains immanently a kind of truth. He is not absolutely certain but he believes profoundly that it is possible to find various sets of a few propositions each such that the propositions of each set are compatible, that the propositions of such a set imply other propositions, and that the latter can be deduced from the former with certainty. That is to say, he believes that there are systems of coherent or consistent propositions, and he regards it his business to discover such systems. Any such system is a branch of mathematics. Any branch contains two sets of ideas (as subject matter, a third set of ideas are used but are not part of the subject matter) and two sets of propositions (as subject matter, a third set being used without being part of the subject): that is, any branch contains a set of ideas that are adopted without definition and a set that are defined in terms of the others; and a set of propositions adopted without proof and called assumptions or principles or postulates or axioms (but not as true or as self-evident) and a set deduced from the former. A system of postulates for a given branch of mathematics — a variety of systems may be found for a same branch — is often called the foundation of that branch. And that is what the layman should think when, as occasionally happens, he meets an allusion to the foundation of the theory of the real variable, or to the foundation of Euclidean geometry or of projective geometry or of *Mengenlehre* or of some other branch of mathematics. The founding, in the sense indicated, of various distinct branches of mathematics is one of the great outcomes of a century of critical activity in the science. It has engaged and still engages the best efforts of men of genius and men of talent. Such activity is commonly described as fundamental. It is very important, but fundamental in a strict sense it is not. For one no sooner examines the foundations that have been found for various mathematical branches and thereby as well as otherwise gains a deep conviction that these branches are constituents of something different from any one of them and different from the mere

sum or collection of all of them than the question super-
venes whether it may not be possible to discover a
foundation for mathematics itself such that the above-
indicated branch foundations would be seen to be, not
fundamental to the science itself, but a genuine part of
the superstructure. That question and the attempt to answer it are fundamental strictly. The question was
forced upon mathematicians not only by developments
within the traditional field of mathematics, but also independently from developments in a field long regarded as
alien to mathematics, namely, the field of symbolic logic.
The emancipation of logic from the yoke of Aristotle very
much resembles the emancipation of geometry from the
bondage of Euclid; and, by its subsequent growth and
diversification, logic, less abundantly perhaps but not less
certainly than geometry, has illustrated the blessings of
freedom. When modern logic began to learn from such
a man as Leibniz (who with the most magnificent expec-
tations devoted much of his life to researches in the subject) the immense advantage of the systematic use of
symbols, it soon appeared that logic could state many of
its propositions in symbolic form, that it could prove
some of these, and that the demonstration could be conducted and expressed in the language of symbols. Evidently such a logic looked like mathematics and acted
like it. Why not call it mathematics? Evidently it differed from mathematics in neither spirit nor form. If it
differed at all, it was in respect of content. But where
was the decree that the content of mathematics should be
restricted to this or that, as number or space? No one
could find it. If traditional mathematics could state and
prove propositions about number and space, about relations of numbers and of space configurations, about
classes of numbers and of geometric entities, modern

logic began to prove propositions about propositions, relations and classes, regardless of whether such propositions, relations and classes have to do with number and space or with no matter what other specific kind of subject. At the same time what was admittedly mathematics was by virtue of its own inner developments transcending its traditional limitations to number and space. The situation was unmistakable: traditional mathematics began to look like a genuine part of logic and no longer like a separate something to which another thing called logic applied. And so modern logicians by their own researches were forced to ask a question, which under a thin disguise is essentially the same as that propounded by the bolder ones among the critical mathematicians, namely, is it not possible to discover for logic a foundation that will at the same time serve as a foundation for mathematics as a whole and thus render unnecessary (and strictly impossible) separate foundations for separate mathematical branches?

It is to answer that great question that Messrs. Whitehead and Russell have written " Principia Mathematica " — a work consisting of four large volumes, the first and second being in hand, the third soon to appear — and the answer is affirmative. The thesis is: it is possible to discover a small number of ideas (to be called primitive ideas) such that all the other ideas in logic (including mathematics) shall be definable in terms of them, and a small number of propositions (to be called primitive propositions) such that all other propositions in logic (including mathematics) can be demonstrated by means of them. Of course, not all ideas can be defined — some must be assumed as a working stock — and those called primitive are so called merely because they are taken without definition; similarly for propositions, not all can

be proved, and those called primitive are so called because they are assumed. It is not contended by the
authors (as it was by Leibniz) that there exist ideas
and propositions that are absolutely primitive in a metaphysical sense or in the nature of things; nor do they
contend that but one sufficient set of primitives (in their
sense of the term) can be discovered. In view of the
immeasurable wealth of ideas and propositions that enter
logic and mathematics, the author's thesis is very imposing; and their work borrows some of its impressiveness from the magnificence of the undertaking. It is
important to observe that the thesis is not a thesis *of*
logic or of mathematics, but is a thesis *about* logic and
mathematics. It can not be proved syllogistically; the
only available method is that by which one proves that
one can jump through a hoop, namely, by actually jump-
ing through it. If the thesis be true, the only way to
establish it as such is to produce the required primitives
and then to show their adequacy by actually erecting
upon them as a basis the superstructure of logic (and
mathematics) to such a point of development that any
competent judge of such architecture will say: " Enough!
I am convinced. You have proved your thesis by
actually performing the deed that the thesis asserts to be
possible."

And such is the method the authors have employed.
The labor involved — or shall we call it austere and
exalted play? — was immense. They had predecessors,
including themselves. Among their earlier works Russell's " Principles of Mathematics " and Whitehead's
" Universal Algebra " are known to many. The related
works of their predecessors and contemporaries, modern
critical mathematicians and modern logicians, Weierstrass, Cantor, Boole, Peano, Schroeder, Peirce and many

others, including their own former selves, had to be digested, assimilated and transcended. All this was done, in the course of more than a score of years; and the work before us is a noble monument to the authors' persistence, energy, acumen and idealism. A people capable of such a work is neither crawling on its belly nor completely saturated with commercialism nor wholly philistine. There are preliminary explanations in ordinary language and summaries and other explanations are given in ordinary language here and there throughout the book, but the work proper is all in symbolic form Theoretically the use of symbols is not necessary. A sufficiently powerful god could have dispensed with them, but unless he were a divine spendthrift, he would not have done so, except perhaps for the reason that whatever is feasible should be done at least once in order to complete the possible history of the world. But whilst the employment of symbols is theoretically dispensable, it is, for man, practically indispensable. Many of the results in the work before us could not have been found without the help of symbols, and even if they could have been thus found, their expression in ordinary speech, besides being often unintelligible, owing to complexity and involution, would have required at least fifteen large volumes instead of four. Fortunately the symbology is both interesting and fairly easy to master. The difficulty inheres in the subject itself.

The initial chapter, devoted to preliminary explanations that any one capable of nice thinking may read with pleasure and profit, is followed by a chapter of 30 pages dealing with " the theory of logical types." Mr. Russell has dealt with the same matter in volume 30 of the *American Journal of Mathematics* (1908). One may or may not judge the theory to be sound or adequate or necessary and yet not fail to find in the chapter setting it forth both an excellent example of analytic and constructive thinking and a worthy model of exposition. The theory, which, however, is recommended by other considerations, originated in a desire to exclude from logic automatically by means of its principles what are called illegitimate totalities and therewith a subtle variety of contradiction and vicious circle fallacy that, owing their presence to the non-exclusion of such totalities, have always infected logic and justified skepticism as to the ultimate soundness of all discourse, however seemingly rigorous. (Such theoretic skepticism may persist anyhow, on other grounds.) Perhaps the most obvious example of an illegitimate totality is the so-called class of all classes. Its illegitimacy may be shown as follows. If A is a class (say that of men) and E is a member of it, we say, E is an A. Now let W be the class of all classes such that no one of them is a member of itself. Then, whatever class x may be, to say that x is a W is equivalent to saying that x is not an x, and hence to say that W is a W is equivalent to saying that W is not a $W!$ Such illegitimate totalities (and the fallacies they breed) are in general exceedingly sly, insinuating themselves under an endless variety of most specious disguises, and that, not only in the theory of classes but also in connection with every species of logical subject-matter, as propositions, relations and propositional functions. As the propositional function — any expression containing a real (as distinguished from an apparent) variable and yielding either non-sense or else a proposition whenever the variable is replaced by a constant term — is the basis of our author's work, their theory of logical types is fundamentally a theory of types of propositional functions.[2] It can not be set forth here nor in fewer pages than the authors have devoted to it. Suffice it to say that the theory presents propositional functions as constituting a summitless hierarchy of types such that the functions of a given type make up a legitimate totality; and that, in the light of the theory, truth and falsehood present themselves each in the form of a systematic ambiguity, the quality of being true (or false) admitting of distinctions in respect of order, level above level, without a summit. When Epimenides, the Cretan, says that all statements of Cretans are false, and you reply that then his statement is false, the significance of " false " here presents two orders or levels; and logic must by its machinery automatically prevent the possibility of confusing them.

Next follows a chapter of 20 pages, which all philosophers, logicians and grammarians ought to study, a chapter treating of Incomplete Symbols wherein by ingenious analysis it is shown that the ubiquitous expressions of the form " the so and so " (the " the " being singular, as " the author of Waverly," " the sine of a," " the Athenian who drank hemlock," etc.) do not of themselves denote anything, though they have contextual significance essential to discourse, essential in particular to the significance of identity, which, in the world of discourse, takes the form of " a is the so and so " and not the form of the triviality, a is a.

After the introduction of 88 pages, we reach the work proper (so far as it is contained in the Volume I.), namely, Part I.: Mathematical Logic. Here enunciation of primitives is followed by series after series of theorems

[2] As to the nature of propositional functions see the earlier chapters of Keyser's *Mathematical Philosophy*, Dutton & Company.

and demonstrations, marching through 578 pages, all matter being clad in symbolic garb, except that the continuity is interrupted here and there by summaries and explanations in ordinary language. Logic it is called and logic it is, the logic of propositions and functions and classes and relations, by far the greatest (not merely the biggest) logic that our planet has produced, so much that is new in matter and in manner; but it is also mathematics, a prolegomenon to the science, yet itself mathematics in the most genuine sense, differing from other parts of the science only in the respects that it surpasses these in fundamentality, generality and precision, and lacks traditionality. Few will read it, but all will feel its effect, for behind it is the urgence and push of a magnificent past: two thousand five hundred years of record and yet longer tradition of human endeavor to think aright.

Owing to the vast number, the great variety and the mechanical delicacy of the symbols employed, errors of type are not entirely avoidable and Volume II. opens with a rather long list of " errata to Volume I." The second volume is composed of three grand divisions: Part III., which deals with cardinal arithmetic; Part IV., which is devoted to what is called relation-arithmetic; and Part V., which treats of series. The theory of types, which is presented in Volume I., is very important in the arithmetic of cardinals, especially in the matter of existence-theorems, and for the convenience of the reader Part III. is prefaced with explanations of how this theory applies to the matter in hand. In the initial section of this part we find the definition and logical properties of cardinal numbers, the definition of cardinal number being the one that is due to Frege, namely, the cardinal number of a class C is the class of all classes similar to

C, where by " similar " is meant that two classes are similar when and only when the elements of either can be associated in a one-to-one way with the elements of the other. This section consists of seven chapters dealing respectively with elementary properties of cardinals; 0 and 1 and 2; cardinals of assigned types; homogeneous cardinals; ascending cardinals; descending cardinals; and cardinals of relational types. Then follows a section treating of addition, multiplication and exponentiation, where the logical muse handles such themes as the arithmetical sum of two classes and of two cardinals; double similarity; the arithmetical sum of a class of classes; the arithmetical product of two classes and of two cardinals; next, of a class of classes; multiplicative classes and arithmetical classes; exponentiation; greater and less. Thus no less than 186 large symbolically compacted pages deal with properties *common* to finite and infinite classes and to the corresponding numbers. Nevertheless finites and infinites do differ in many important respects, and as many as 116 pages are required to present such differences under such captions as arithmetical substitition and uniform formal numbers; subtraction; inductive cardinals; intervals; progressions; Aleph null, $\aleph_0$; reflexive classes and cardinals; the axiom of infinity; and typically indefinite inductive cardinals.

As indicating the fundamental character of the " Principia " it is noteworthy that the arithmetic of relations is not begun earlier than page 301 of the second huge volume. In this division the subject of thought is relations including relations between relations. If R_1 and R_2 are two relations and if F_1 and F_2 are their respective fields (composed of the things between which the relations subsist), it may happen that F_1 and F_2 can be so correlated that, if any two terms of F_1 have the relation

R_1, their correlates in F_2 have the relation R_2, and *vice versa*. If such is the case, R_1 and R_2 are said to be *like* or to be *ordinarily similar*. Likeness of relations is analogous to similarity of classes, and, as cardinal number of classes is defined by means of class similarity, so relation-number of relations is defined by means of relation likeness. And 209 pages are devoted to the fundamentals of relation arithmetic, the chief headings of the treatment being ordinal similarity and relation-numbers; in-internal transformation of a relation; ordinal similarity; definition and elementary properties of relation-numbers; the relation-numbers, 0_r, 2_r and 1_s; relation-numbers of assigned types; homogeneous relation-numbers; addition of relations and the product of two relations; the sum of two relations; additions of a term to a relation; the sum of the relations of a field; relations of mutually exclusive relations; double likeness; relations of relations of couples; the product of two relations; the multiplication and exponentiation of relations; and so on.

The last 259 pages of the volume deal with series. A large initial section is concerned with such properties as are common to all series whatsoever. From this exceedingly high and tenuous atmosphere, the reader is conducted to the level of sections, segments, stretches and derivatives of series. The volume closes with 58 pages devoted to convergence, and the limits of functions.

To judge the " Principia," as some are wont to do, as an attempt to furnish methods for developing existing branches of mathematics, is manifestly unfair; for it is no such attempt. It is an attempt to show that the entire body of mathematical doctrine is deducible from a small number of assumed ideas and propositions. As such it is a most important contribution to the theory of the unity of mathematics and of the compendence of knowledge in general. As a work of constructive criticism it has never been surpassed. To every one and especially to philosophers and men of natural science, it is an amazing revelation of how the familiar terms with which they deal plunge their roots far into the darkness beneath the surface of common sense. It is a noble monument to the critical spirit of science and to the idealism of our time.

CONCERNING MULTIPLE INTERPRETA-TIONS OF POSTULATE SYSTEMS AND THE "EXISTENCE" OF HYPER-SPACE [1]

WHAT do we mean when we speak of n-dimensional space and n-dimensional geometry, where n is greater than 3? The question refers to talk about space and geometry that are n-dimensional in *points*, for ordinary space, as is well known, is 4-dimensional in lines, 4-dimensional in spheres, 5-dimensional in flat line-pencils, 6-dimensional in circles, etc., and there is naturally no mystery involved in speaking of these latter varieties of multi-dimensional manifolds and their geometries, no matter how high the dimensionality may be. No mystery for the reason that in these geometries everything lies within the domain of intuition in the same sense in which everything in ordinary (point) geometry lies in that domain. In other words, these n-dimensional geometries are nothing but theories or geometries of *ordinary* space, that arise when we take for element, not the point, but some other entity, as the line or the sphere, . . . whose determination in ordinary space requires more than 3 independent data. Of these varieties of n-dimensional geometry, the inventor was Julius Plücker (d. 1868), but Plücker declined to concern himself with spaces and geometries of more than four dimensions in *points*.

[1] Printed in *The Journal of Philosophy, Psychology and Scientific Method*, May 8, 1913.

Since Plücker's time, however, such hyper-theories of points have invaded not only almost every branch of pure mathematics, but also — strangely enough — certain branches of physical science, as, for example, the kinetic theory of gases. As to the manner of this latter invasion a hint may be instructive. Given N gas molecules enclosed, say, in a sphere. These molecules are, it is supposed, flying about hither and thither, all of them in motion. Each of them depends on six coordinates, x, y, z, u, v, w, where x, y, z, are the usual positional coordinates of the molecule regarded as a point in ordinary space, and u, v, w are the components of the molecule's velocity along the three coordinate axes. Knowing the six things about a given molecule, we know where it is and the direction and rate of its going. The N molecules making up the gas depend on $6N$ coordinates. At any instant these have definite values. These values together define the " state " of the gas at that instant. Now these $6N$ values are said to determine a point in space of $6N$ dimensions. Thus is set up a one-one correspondence between such points and the varying gas states. As the state of the gas changes, the corresponding point generates a locus in the space of $6N$ dimensions. In this way the behavior or history of the gas gets geometrically represented by loci in the hyperspace in question.

Is such geometric n-dimensional phraseology merely a geometric way of speaking about non-spatial things? Even if there exists a space, S_n, one *may* employ the language appropriate to the geometry of the space without having the slightest reference to it, and, indeed, without knowing or even enquiring whether it exists. This use of geometric speech in discourse about non-spatial things is not only possible, but in fact very common. An easily accessible example of it may be found in Bôcher [2] where, in speaking of a set of values of n independent variables as a point in space of n dimensions the reader is told that the author's use of geometric language for the expression of algebraic facts is due to certain advantages of that language compared with the language of algebra or of analysis; he is told that the geometric terms will be employed " in a wholly conventional algebraic sense " and that " we do not propose even to raise the question whether in any geometric sense there is such a thing as space of more than three dimensions."

It is held by many, including perhaps the majority of mathematicians, that there are no hyperspaces of points and that n-dimensional geometries are, rightly speaking, not geometries at all, but that the facts dealt with in such so-called geometries are nothing but algebraic or analytic or numeric facts expressed in geometric language. If this opinion be correct, then the extensive and growing application of geometric language to analytical theories of higher dimensionality indicates a high superiority of geometric over analytic speech, and it becomes a problem for psychology to ascertain whether the mentioned superiority is adequate to explain the phenomenon in question and, if it be adequate, to show wherein the superiority resides.

No doubt geometric language has a kind of esthetic value that is lacking in the speech of analysis, for the former, being transfused with the rich reminiscences of sensibility, constantly awakens a delightful sense, as thinking proceeds, of the colors, forms, and motions of the sensuous world. This is an emotional value. No doubt, too, geometric language has, in its distinctive conciseness, an economic superiority, as when, for example, one speaks of the points of the 4-dimensional sphere,

[2] " Introduction to Higher Algebra," page 9.

$x^2 + y^2 + z^2 + w^2 = r^2$, instead of speaking of the various systems of values of the variables x, y, z, w that satisfy the equation $x^2 + \ldots = r^2$. Additional advantages of geometric over analytic speech are brought to light in the following remarks by Poincaré in his address, " L'Avenir des Mathématiques " (1908):

" Un grand avantage de la géométrie, c'est precisément que les sens y peuvent venir au secours de l'intelligence, et aident à deviner la route à suivre, et bien des esprits préferènt ramener les problèmes d'analyse à la forme géométrique. Malheureusement nos sens ne peuvent nous mener bien loin, et ils nous faussent compagnie dès que nous voulons nous envoler en dehors des trois dimensions classiques. Est-ce à dire que, sortis de ce domaine restreint où ils semblent vouloir nous enfermer, nous ne devons plus compter que sur l'analyse pure et que toute géométrie à plus de trois dimensions est vaine et sans objet? Dans la génération qui nous a précédés, les plus grands maîtres auraient répondu ' oui '; nous sommes aujourd'hui tellement familiarisés avec cette notion que nous pouvons en parler, même dans un cours d'université, sans provoquer trop d'étonnement.

" Mais à quoi peut-elle servir? Il est aisé de le voir: elle nous donne d'abord un langage très commode, qui exprime en termes très concis ce que le langage analytique ordinaire dirait en phrases prolixes. De plus, ce langage nous fait nommer du même nom ce qui se ressemble et affirme des analogies qu'il ne nous permet plus d'oublier. Il nous permet donc cenore de nous diriger dans cet espace qui est trop grand pour nous et que nous ne pouvons voir, en nous rappelant sans cesse l'espace visible qui n'en est qu'une image imparfaite sans doute, mais que en est encore une image. Ici encore, comme dans tous le exemples précédents, c'est l'analogie avec ce qui est simple qui nous permet de comprendre ce qui est complexe."

The question of determining the comparative advantages and disadvantages of the languages of geometry and analysis is a very difficult one. It is evidently in the main a psychological problem. It appears that no serious and systematic attempt has ever been made to solve it. Here, it seems, is an inviting opportunity for a properly qualified psychologist, it being understood that proper qualification would include a familiar knowledge of the languages in question. The interest and manifold utility of such a study are obvious. In the course of such an investigation it would probably be found that the superiority of geometric over analytic speech is alone sufficient to account for the extensive and rapidly increasing literature of what is called n-dimensional geometry and that, in order to account for the rise of such literature, it is therefore not necessary to suppose the existence of n-dimensional spaces, S_n, the facts dealt with in the literature being, it could be supposed, nothing but analytic facts expressed in geometric language.

If such a result were found, would it follow that S_n does not exist and that consequently n-dimensional geometry must be nothing but analysis in geometric garb? The answer is, no; for we may and we often do assign an adequate cause of a phenomenon or event without assigning the actual cause; and so the possibility would remain that n-dimensional geometry has an appropriate object or subject, namely, a space S_n, which, though without sensuous existence, yet has every kind of existence that may warrantably be attributed to ordinary geometric space, S_3. For this last, though it is imitated by (or imitates) sensible space, as an ideal model or pattern is imitated by (or imitates) an imperfect copy, it is not identical with it. S_3 is not tactile space, nor visual space, nor that of muscular sensation, nor the space of any other sense, nor of all the senses — it is a conceptual space; and whether there are or are not spaces S_4, S_5, etc., which have every sort of existence rightly attributable to ordinary geometric spaces, S_3, and which differ from the latter only in the accident of dimensionality and in the further accident that S_3 appears in the rôle of an ideal prototype for an actual sensible space, whilst S_4, S_5, etc., do not present such an appearance, — that is the question which remains for consideration.

A friend called at my study, and, finding me at work, asked, " What are you doing? "   My reply was: " I am trying to tell how a world which probably does not exist would look if it did."   I had been at work on a chapter of what is called 4-dimensional geometry. The incident occurred ten years ago. The reply to my friend no longer represents my conviction. Subsequent reflection has convinced me that a space, S_n, of four or more dimensions has every kind of existence that may be rightly ascribed to the space, S_3, of ordinary geometry.

The following paragraphs present — merely in outline, for space is lacking for a minute presentation — the considerations that have led me to the conclusion above stated.

Let sensible space be denoted by sS_3. We know that sS_3 is discontinuous (in the mathematical sense of the term) and that it is irrational. By saying that it is irrational I mean what common experience as well as the results of experimental psychology prove: that three sensible extensions of a same type, let us for definiteness say three sensible lengths, l_1, l_2, l_3, may be such that

(1) $l_1 = l_2, l_2 = l_3, l_1 \neq l_3.$

Because sS_3 is thus irrational, because it is radically infected with such contradictions as (1), this space is not, and can not be, the subject or object of geometry, for geometry is rational; it does not admit three such extensions as those in (1). Not only do such contradictions as (1) render sS_3 impossible as a subject or object of geometry, but, when encountered, they produce intolerable intellectual pain — nay, if they could not in somewise be transcended or overcome, they would produce intellectual death, for, unless the law of non-contradiction be preserved, concatentative thinking, the life of intellect, must cease. In case of intellect we may say that its struggle for existence is a struggle against contradictions. But mere existence is not the characteristic aim or aspiration of intellect. Its aim, its aspiration, its joy, is compatibility. Indeed, intellect seems to be controlled by two forces, a *vis a tergo* and *a vis a fronte:* it is driven by discord and drawn by concord. Intellect is a perpetual suitor, the object of the suit being harmony, the beautiful daughter of the muses. Its perpetual enemy is the immortal demon of discord, ever being overcome, but never vanquished.

The victory of intellect over the characteristic contradictions inherent in sS_3 is won through what we call conception. That is to say that either we find or else we create another kind of space which, in order to distinguish it nominally and symbolically from sS_3, we may call conceptual space, and denote by cS_3. Unlike sS_3, cS_3 is mathematically continuous and it is rational. Like sS_3, cS_3 is extended, it has room, but the room and the extensions are not sensible, they are conceptual; and these extensions are such that, if l_1, l_2, l_3 be three amounts of a given type of extension, as length, say, and if $l_1 = l_2$ and $l_2 = l_3$, then $l_1 = l_3$. The space cS_3, whether we regard it as found by the intellect or as created by it, is the subject or object of geometry. The current vulgar confusion of sS_3 and cS_3 is doubtless due to the fact that the former imitates the latter, or the latter the former, as a sensible thing imitates its ideal, or as an ideal (of a sensible thing) may be said to imitate that thing; for it is precisely such alternative or mutual imitation that enables us in a measure to control the sensible world through its conceptual counterpart; and so the exigencies of practical affairs and the fact that reciprocally imitating things each reminds us of the other coöperate to cause the sensible and the ideal, the perceptual and the conceptual, to mingle constantly and to become confused in that part of our mental life that belongs to the sensible and the conceptual worlds of three dimensions. Nevertheless, it is a fact to be borne in mind that cS_3 is a subject or object of geometry and that sS_3 is not.

Now, in order to construct the geometry in question, we start with a suitable system of postulates or axioms expressing certain relations among what are called the elements of cS_3. These postulates, together with such propositions as are deducible from them, constitute the geometry of cS_3. I shall call it *pure* geometry, for a reason to be given later, and shall denote it by pG_3. For definiteness let us refer to the famous and familiar postulates of Hilbert. Any other system would do as well. In the Hilbert system, the elements are called points, lines, and planes. It is customary and just to point out that the terms point, line, and plane are not defined, and in critical commentary it is customary to add:

(A) That, consequently, these terms may be taken to be the names of any things whatsoever with the single restriction that the things must satisfy the relations stated by the postulates;

(B) That, when some admissible or possible interpretation I has been given to the element-names, the postulates P together with their deducible consequences C constitute a definite theory or doctrine $D;$

(C) That replacing I by a different interpretation I' produces no change whatever in $D;$

(D) That this invariant D is Euclidean geometry of three dimensions; and

(E) That, if we are to speak of D as a theory or geometry of a space, this space is nothing but the ensemble of any kind of things that may serve for an interpretation of P.

That the view expressed in that so-called " critical commentary " does not agree with common sense or with traditional usage is obvious. That it will not bear critical reflection can, I believe, be made evident. Let us examine it a little. In order to avoid the prejudicial associations of the terms point, line, and plane, we may replace them by the terms " roint," " rine," and " rane," so that the first postulate, or axiom, as Hilbert calls it, will read: *Two distinct roints always completely determine a rine.* Or, better still, we may replace them by the symbols e_1, e_2, e_3, so that the reading will be: *Two distinct e_1's always completely determine an e_2*; and similarly for the remaining postulates.

We will suppose the phrasing of (A), (B), (C), (D), (E), slightly changed to agree with the indicated new phrasing of the postulates.

It seems very probable that there are no termless relations, *i.e.*, relations that do not relate. It seems very probable that a relation to be a relation must be something actually connecting or subsisting between at least two things or terms. A postulate expressing a relation having terms is at all events ostensibly a statement about the terms, and so it would seem that, if the relation be supposed to be termless, the statement ceases to be a statement about something and, in so ceasing, ceases to be a statement that is true or else is false. In discourse, it is true, there is frequent seeming evidence that relations are often thought of as termless, as when, for example, we speak of " a relation *and* its terms "; but then we speak also of a neckless fiddle without intending to imply by such locution that there can be a fiddle without a neck. As, however, we do not wish the validity of the following criticism to depend on the denial of the possibility of termless relations, the discussion will be conducted in turn under each of the alternative hypotheses: (h_1) There are termless relations; (h_2) There are no termless relations. We will begin with

Hypothesis h_2

To (A) we make no objection.

Let us now suppose given to P some definite interpretation I. Let us grant that we now have a definite doctrine D, consisting of P and C. Either the things which in I the e's denote have or they have not content, character, or meaning, m, in excess of the fact that they satisfy P.

(1) Suppose they have not an excessive meaning m. Denote the interpretation by I_1 and the doctrine by D_1. This D_1 is a queer doctrine. We may ask: what does D_1 relate or refer to? That is, what is it a doctrine of or about? The question seems to admit of no intelligent or intelligible answer. For if the doctrine is about something, it is, it seems natural to say, a doctrine about the I_1-things (denoted by the e's); but, by (1), these I_1-things can not be characterized or indicated otherwise than by the fact of their satisfying P; and so it appears

that such attempted natural answer is reducible and equivalent to saying (a) that the doctrine D_1 is about the things which it is about. In order not to be thus defeated, one might try to give an informing answer by saying that D_1 is a doctrine, not about the I_1-things, *i.e.*, not about terms of relations, but about the relations themselves. Such an answer is suspicious on account of its unnaturalness, and it is unnatural because the propositions of D_1 wear the appearance of talking explicitly, not about relations, but about terms of relations. Moreover, the answer is not an informing one unless the relations that the doctrine D_1 is alleged to be about can be characterized otherwise than by the fact of their being satisfied by the I_1-things, for, if they can not be otherwise characterized, evidently by (1) the answer reduces to a form essentially like that of (a). May not one escape by saying that the relations which D_1 is alleged to be a doctrine about are just the relations expressed by the propositions in D_1? Does this attempted characterization make the answer in question an informing one? If D_1 is a doctrine about the relations expressed by its propositions, then D_1 says or teaches something about these relations, for every doctrine, if it be about something, must teach or say something about that which it is about. In the case supposed, what does D_1 teach about the relations? Nothing except that they are satisfied by the I_1-things. In other words, what D_1 teaches about the relations expressed by its propositions is, by (1), that these are satisfied by things that satisfy them — a not very nutritious lesson. It is possible to make a yet further attempt so to indicate the relations as to render the answer, that the doctrine D_1 is about relations, an informing one. It is known that P may receive an interpretation I' different from I_1 in that the I'-things do not satisfy (1), but have an excessive content, character, or meaning m. May we not give the required indication of the relations that D_1 is said to be a doctrine about by saying that they are relations satisfied by the I'-things, the presence of the m involved making the indication genuine or effective? It seems so at first. But if again we ask what D_1 teaches about the relations thus indicated, we are led into the same difficulty as above. Moreover, when we ask what D_1 is a doctrine about we expect an answer in terms in somewise mentioned or intimated in the D_1 discourse, whilst in the case in hand the required indication has depended on m, a thing expressly excluded from the D_1 discourse by (1).

So, I repeat, D_1 is a queer doctrine.

It must be added that if there be an interpretation I_1, it is unique of its kind; for if I_1' were an interpretation satisfying (1), the I_1'-things would have no excessive meaning m; hence they would be simply the I_1-things, and I_1 and I_1' would be merely two symbols for a same interpretation.

Accordingly, if there were an interpretation I_1, but no other, i.e., no interpretation I in which the I-things did not satisfy (1), then (C) would be pointless; by (D), D_1 would be Euclidean geometry of three dimensions; and, by (E), Euclidean space, if we wished to speak of D_1 as a geometry of a space, would be the ensemble of the I_1-things; but, if we wished to characterize the I_1-things, the elements of Euclidean space, we could only say that they are the things satisfying certain relations, and, if we wish to indicate what relations, we could only say, the relations satisfied by those things: a very handsome circle.

In the following it will be seen that we are in fact not imprisoned within that circle.

(2) Suppose the I-things of the above-assumed interpretation I do not satisfy (1), but have an excessive meaning m. (It is known that such an I is possible, an example being found by taking for an e_1 any ordered triad of real numbers (x, y, z); for an e_2 the ensemble of triads satisfying any two distinct equations,

$$A_1x + B_1y + C_1z + D_1 = 0,\ A_2 + B_2y + C_2z + D_2 = 0,$$

in neither of which the coefficients are all of them zero; and for an e_3 the ensemble of triads satisfying any one such equation; the presence of m being evident in countless facts such as the fact, for example, that an e_1 is composed of numbers studied by school-boys or useful in trade without regard to their ordered triadic relationship.) Denote the assumed definite interpretation I by I_2 to remind us that it satisfies (2), and denote the corresponding doctrine by D_2. It is immediately evident that there is an interpretation I_1 and hence a doctrine D_1, for to obtain I$_1$ it is sufficient to abstract from the m of the I_2-things and to take the abstracts (which plainly satisfy (1)) for I$_1$-things.

Are D_1 and D_2 but two different symbols for one and the same doctrine, as asserted by (C)? Evidently not. For, in respect of D_2, we can give an informing answer to the question, what is D_2 a doctrine about? Owing to the presence of the m in the I_2-things, the answer will be an informing one whether it be the natural answer that D_2 is a doctrine about I_2-things, or one of the less natural answers, that D_2 is about the relations having the I_2-things for terms, that D_2 is about the relations expressed by its propositions; whilst, as we have seen, owing to the absence of m in the I_1-things no such answers were, in respect of D_1, informing answers.

Can not (C) be saved by refusing to admit that there

is an interpretation I_1, and so refusing to admit that there is a D_1? If there is no I_1 and hence no D_1, then (C) is pointless unless there is an I_2' and so a D_2' in which the I_2'-things have an m' different from the m of the I_2-things. But if there is an I_2' thus different from I_2, then obviously D_2' and D_2 are, contrary to (C), different doctrines, for they are respectively doctrines about the I_2-things and the I_2'-things, and these thing-systems are different by virtue of the difference of m and m'. Now, it is known that there are two such differing interpretations I_2 and I_2'. For we may suppose I_2 to be the possible interpretation indicated in the above parenthesis. And for I_2' we may take for e_1 any ordered triad of real numbers, except a specified triad (a, b, c), and including $(\infty, \infty, \infty,)$; for e_2 the ensemble of triads, except (a, b, c), that satisfy any pair of equations,

$$A_1(x^2 + y^2 + z^2) + 2B_1(x - a) + 2C_1(y - b)$$
$$+ 2D_1(z - c) - A_1(a^2 + b^2 + c^2) = 0,$$
$$A_2(x^2 + y^2 + z^2) + 2B_2(x - a) + 2C_2(y - b)$$
$$+ 2D_2(z - c) - A_2(a^2 + b^2 + c^2) = 0;$$

and for e_3 the ensemble satisfying any one such equation. Just as when we compared D_1 and D_2, so here the conclusion is, that (C) is not valid.

As a matter of fact mathematicians know that there are possible infinitely many different interpretations of P. It follows from the foregoing that there are correspondingly many different doctrines. For the sake of completeness we may include D_1 among these, although, for the purpose of answering a hypothetical objection, we momentarily supposed D_1 to be disputable or inadmissible.

Which one of the D's is (or should be called) Euclidean geometry of three dimensions? I say which "one"?

For, as no two are identical, it would be willful courting of ambiguity to allow that two or more of them should be so denominated. Which *one,* then? Evidently not one of the numerical ones, such, for example, as the two above specified. For who has ever *really* believed that a point, for example, *is* a triad of numbers? We know that the Greeks did arrive at geometry; we know that they did not arrive at it through numbers; and we know that, in their thought, points were not number triads, nor were planes and lines, for them, certain ensembles of such triads. The confusion, if anybody ever was really thus confused, is due to the *modern* discovery that number triads and certain ensembles of them happen to satisfy the same relations as the Greeks found to be satisfied by what they called points, lines, and planes. There is really no excuse for the confusion, for, if Smith is taller than Brown, and yonder oak is taller than yonder beech, it obviously does not follow that Smith is the oak and Brown the beech.

Evidently Euclidean geometry of three dimensions is that particular *D* for which the *I*-things are points, lines, and planes. Here it is certain to be asked: What, then, are points, lines, and planes? And the asker will mean to imply that, in order to maintain the proposition, it is necessary to define these terms. The proper reply is that it is not necessary to *define* them. All that can be reasonably required is that they be indicated, pointed out, sufficiently described for purposes of recognition, for what we desire is to be able to say or to recognize what Euclidean geometry is about. To the question one might, not foolishly, reply that the terms in question denote things that you and I, if we have been disciplined in geometry, converse understandingly about when we converse about geometry, though neither of us is able to say with *absolute precision* what the terms mean. For who does not know that it is possible to write an intelligent and intelligible discourse about cats, for example, without being able to tell (for who can tell?) precisely what a cat is? And if it be asked what the discourse is about, who does not know that it is an informing answer to say that it is about cats? It is informing because the term cat has an excessive meaning, a meaning beyond that of satisfying the propositions (or relations) of the discourse.

Just here it is well worth while to point out an important lesson in the procedure of Euclid. Against Euclid it is often held as a reproach that he attempted to define the element-names, point, line, and plane, since no definitions of them could render any logical service, that is, in the strictly deductive part of the discourse. But to render no logical service is not to render no service. And the lesson is that the definitions in question, which it were perhaps better to call descriptions, do render an extralogical service. They render such service not only in guiding the imagination in the matter of invention, but also in serving to indicate, with a goodly degree of success, the excessive meaning m of the elements denoted by the terms in question and in thus serving to make known what it is that the deductive part of the discourse is about. One should not forget that no discourse, no doctrine, not even so-called pure logic itself, is exclusively deductive, for in any doctrine there is reference, implicit or explicit, to something extradeductive or extralogical, reference, that is, to something which the doctrine is about.

Are the three Euclidean " definitions," thus viewed as descriptions, sufficient or adequate to the service that they are here viewed as rendering? If by sufficient or adequate be meant exhaustive, the answer is, of course, no. For we may confidently say that no possible description, that is, no description involving only a finite number of words, can exhaust the meaning of a system of terms except, possibly, in the special case where these have no meaning beyond what they must have in order merely to satisfy a finite number of postulates. But exhaustive is not what is meant by adequate. To employ a previous illustration, it is not necessary to give or to attempt an exhaustive description of " cat " in order to tell adequately what it is that a discourse ostensibly about cats is ostensibly about. It is a question of intent. A description is nearly, if not quite, adequate if it enables us to avoid thinking that terms are intended to denote what they are not intended to denote. And, whilst we may not admit that the three Euclidean " descriptions " are the best that can be invented for the purpose, yet we must allow that they have long served the end in question pretty effectively and that they are qualified to continue such service. They have been and they are good enough, for example, to save us from thinking that the things which in geometry have been denoted by the terms, points, lines, and planes, are identical with number triads, etc. The open secret of their thus saving us is no doubt in their causing us to think of points, lines, and planes in terms of, or in essential connection with, what we know as *extension*, whilst numbers and number ensembles are not things naturally so conceived. For evidently the notions of " length " and " breadth " involved in the Euclidean " descriptions " are not *metric* in meaning; they do not signify definite or numeric quantities or amounts of something (as when we say the length of this or that thing is so and so much); but plainly they are generic notions connoting *extension*. It is safe to say that a mind devoid of the concept or the sense of extension could not know what things the " descriptions " aim at describing. It is true that Euclid's " description " of a point as " that which has no part " implies a *denial* of extension, but the denial is one of *extension,* and, in its contextual atmosphere, it is *felt* to be essential to an adequate indication of what is meant by point. On the other hand, if one were (and how unnatural it would be!) to describe an ordered triad of numbers as " that which has no part," it would be immediately necessary to explain away the seeming falsity of the description by saying that the triad is not the ordered multiplicity (of three numbers) *as* a multiplicity, but is merely the *uniphase* of the multiplicity, and that it is this uniphase which has no part. If, next, we were to say that thus extension is denied to the uniphase, the statement, though true, would be *felt* to be inessential to an adequate indication of what is meant by a triad of numbers. Such felt difference is alone sufficient to make any one pause who is disposed to adopt the current creed that a point is nothing but an ordered triad of numbers. It is not contended that a point is *composed* of extension; the contention is that point and extension are so connected that a mind devoid of the latter notion would be devoid of the former, just as a mind devoid of the notion of variable or variation would be devoid of the notion of constant, though a constant is not a thing *consisting* of variation; just as the notion of limit would not be intelligible except for the notion of something that may have a limit, though the limit is not *composed* of it; and just as an instant, which is not *composed* of time, would not be intelligible except for the notion of time.

In a discussion of such matters it is foolish and futile to talk about " proofs." The question, as said, is one of intent; it is a question of self-veracity, of getting aware of and owning what it is that we mean by the terms and symbols of our discourse. If, despite the Euclidean " descriptions " and despite any and all others that may supplement or supplace them, one fails to see that *extension* is essentially involved in the meaning that the terms points, lines, and planes, are intended to have, the failure will be because " as the eyes of bats are to the blaze of day, so is the reason in our soul to the things which are by nature most evident of all." Nothing is more evident than that there is something that is called extension. We have but to open our eyes to get aware that we are beholding an *expanse,* something extended. We see things *as* extended: things *as* extended are revealed to the tactile sense; a region or room involving extension is a datum of the muscular sensations connected with our bodily movements; and so on. So much is certain. But it is said and rightly said that these are *sensible* things; that the extension they are revealed as having is *sensible* extension; that these sensibles are infected with contradiction, above noted, revealed in common experience, and confirmed by the psychophysical law of Weber and Fechner; that geometry is free from contradiction, that, therefore, geometry is not a doctrine about these sensibles; that among these sensibles are not the things which in geometry are denoted by the terms point, line, and plane; and that, if these terms imply or connote extension, as asserted, this extension is not sensible extension. Granted. The " connoted extension " is not sensible, it is conceptual. How know, however, that there is conceptual extension? The answer is, by arriving at it. (We need not here debate whether such " arriving " is best called creating or is best called finding.) But how does the mind arrive at it? By doing

certain things to the sensibles, the raw materials of
mental architecture. What things? An exhaustive answer is unnecessary — perhaps impossible. The things
are of two sorts: the mind gives to the sensibles; it takes
away from them. Consider for example a sensible line.
From it the conceptualizing intellect takes away (abstracts from, disregards) certain things that the sensible
in question has or may have, as color, weight, tempera-
ture, etc., including *part* of the extension, thus, I mean,
narrowing and thinning away all breadth and thickness.
What of the extension called length? Have the narrowing and thinning taken it away? It was not so intended,
the opposite was intended. Yet no sensible length (extension) remains. Does the narrowing and thinning involve
shortening? We are absolutely certain that it does not.
What, then, is it that has happened? Evidently that, by
the indicated taking away, the mind has arrived at insensible length, one kind of insensible extension, that is,
at conceptual length, one kind of conceptual extension.
A stretch, we are sure, remains, but it is not a sensible
stretch. The extension thus arrived at is yet not the
extension connoted by or involved in the things that
geometry is about, for in the taking-away process of
arriving at it there is nothing to disinfect it of the contradictions inherent in the sensible with which we started.
It remains, then, to follow the indicated process of taking
away by a process of giving, that is to say, it remains
to endow the conceptual extension (arrived at) with
continuity so as to render it free from the mentioned
contradictions. This done, the kind of extension meant
in ordinary geometry or ordinary geometric space is
arrived at. Such is, in kind, the conceptual extension
that, it is here held, is essential to what the geometric
terms, point, line, plane, are intended to mean. Without

further talk we may say that such extension is essential in the conceptual space that, we may say, ordinary Euclidean geometry is about in being about the elements of the space.

If we denote this conceptual space by cS_3 to distinguish it from (non-geometrizable) sensible space sS_3, then the geometry of cS_3, if constructed by means of postulates P making no indispensable use of algebraic analysis, may be called pure geometry, pG_3. If, as in the Cartesian method, we use ordered number triads, etc., as we may use them, not to be points, etc., but to represent points, etc., then we get analytical geometry, aG_3, of cS_3. On the other hand, if, as we may, we interpret the P by allowing the I-things to be number triads, etc., as above indicated, the resulting doctrine is, not geometry, but a pure algebra or analysis, pA_3. If we use points, etc., not to be, but to represent, number triads, etc., and so employ geometric language in constructing pA_3, we get by this kind of anti-Cartesian procedure, not a geometry, but geometrical analysis, gA_3.

Hypothesis h_1

It is unnecessary to say anything and is not worth while to say much under this hypothesis. For if the e's in P do not denote something, then as the relations (if there be any) are termless, the doctrine D (if there be one) is not about anything, unless about the relations, but about these it says nothing, for, if it says aught about them, what it says is that they are satisfied by certain terms whose presence in the discourse is excluded by h_1. We may profitably say, however, that, in the case supposed where the e's do not denote something but are merely uninterpreted variables ready, so to speak, to denote something — in this case we may say that, though there is no doctrine *D*, there is a *doctrinal function*,[8] $\triangle$ (e_1, e_2, e_3). Then we should add that the doctrines that do arise from actualized possible interpretations of the *e*'s are so many *values* of $\triangle$. This function $\triangle$, if we give some warning mark as $\triangle'$ to its symbol, may be further conveniently employed in talking about an ambiguous one of the doctrines in question, *i.e.*, about " any value," an ambiguous value, of the function. As above argued, these values, these doctrines are identical in *form*, they are *isomorphic*, all of them having the form of $\triangle$, but no two of them are the same in respect of content, reference, or meaning. In this conclusion, analysis, happily, agrees with traditional usage, intuition, and common sense.

CONCLUDING CONSIDERATIONS

We are, I believe, now prepared to answer definitively the long-vexed question: *What, if any, sort of existence have point spaces of four or more dimensions?*

As we have seen, the conceptual space cS_3 of ordinary geometry is an affair involving extension; it is a triply extended conceptual spread or expanse: three independent linear extensions in it may be chosen; these suffice to determine all the others. So much is as certain as anything can be. It is equally certain that we can, for we do without meeting contradiction, by means of postulates or otherwise, conceive (not perceive or imagine) a quadruply extended spread or expanse, one, that is, in which it is possible to choose four independent linear extensions, and then by reference to these to determine all

---

[8] The notion I have thus named here is formally introduced, explained and appraised in my *Mathematical Philosophy* (Dutton and Company).

the rest. There is not the slightest difference in kind among the four independents and not the slightest difference between any three of these and the three of cS_3. The spread or expanse thus set up is a cS_4; like cS_3, it is purely conceptual; the extension it involves is, in kind, identical with that of cS_3; it contains spreads of the type of cS_3 as elements just exactly as a cS_3 contains planes or spreads of type cS_2 as elements; it differs not at all from cS_3 except in being one degree higher in respect of dimensionality. In a word, cS_4 (and, of course, cS_5, and so on) has the same kind of existence as cS_3. It is true that cS_3 is " imitated " by our sensible space sS_3, whilst there is no sS_4 thus imitating cS_4. But this writing is not intended for one who is capable of thinking that the mentioned sensible imitation or imitability of cS_3 confers upon the latter a new or peculiar kind of existence.

But one thing remains to be said, and it is important. If one denies that cS_3 has the conceptually extensional existence, above alleged, then, of course, the denial extends also to cS_4, and the two spaces are, in respect to existence, still on a level. If the denier then asserts, and such is the alternative, that cS_3 is only the ensemble of number triads, etc., as above explained, then, if he be right, cS_4 is only, but equally, the ensemble of ordered quatrains, etc., of numbers. Here, again, cS_3 and cS_4 have precisely the same kind of existence. The conclusion is that *hyperspaces have every kind of existence that may be warrantably attributed to the space of ordinary geometry.*[4]

[4] See the lecture devoted to Hyperspace in Keyser's *Mathematical Philosophy.*

MATHEMATICAL PRODUCTIVITY IN THE UNITED STATES [1]

BOTH on its own account and in its relation to the general question of research, this subject is naturally interesting to the specialists immediately concerned; and it seems a happy augury that not long ago several western college and university presidents, in convention, considered the problem of how to secure that officers of instruction shall become, in addition, investigators and producers. A complete solution will be found when, and only when, the nature and importance of the problem shall be appreciated, not only by scientific specialists and university presidents, but by educators and the educated public in general, and this condition will be satisfied in proportion as the interdependence of all grades and varieties of educational and scientific activity shall come to be generally understood, and especially in proportion as we learn to value the things of mind, not merely for their utility, but for their spiritual worth, and to seek, as a community, in addition to comfort and happiness, the glory of the sublimer forms of knowledge and intellectual achievement.

Except when the contrary may be indicated or clearly implied, the discussion will confine itself to pure mathematics as distinguished from applied mathematics, such as mechanics and mathematical physics.

And first as to the significance of terms. According to the usage that has long prevailed among foreign

[1] Printed in the *Educational Review*, November, 1902.

mathematicians, and which, during the last quarter of a century, has come to prevail also in this country, the term mathematical productivity is restricted to discovery, successful research, extension in some sense of the boundaries of mathematical knowledge; and such productive activity includes and ranges thru the establishment of important new theorems, the critical grounding of classical doctrines, the discovery or invention of new methods of attack, and, in its highest form, the opening and exploration of new domains.

Not only does the term productivity now signify here what it signifies abroad, but the prevailing standards in the United States agree with those of Europe. It is not meant that the best work in this country is yet equal to the *very* best of the European, nor that the averages coincide, but that the Americans judge home and foreign products by the same canons of value, and that these are as rigorous as the French, German, or British rules of criticism.

Time was when productivity meant, in the United States, the writing and publishing of college text-books in algebra, geometry, trigonometry, analytical geometry and the calculus, not to mention arithmetic. That time has gone by. At present the term neither signifies such work nor, except in rare instances, includes it. With reverence for the olden time when the college professor, especially in comparison with the average of his successors of the present time, was apt to be a man of general attainments and diversified learning, it may be said that, judged by modern standards of specialized scholarship, the special attainments of American mathematicians previous to a generation ago, except in the case of a few illustrious men, were exceedingly meagre — a fact which, as it could hardly have been suspected owing to their isolation by the mathematicians themselves, was even less known to their colleagues in other branches of learning or to the educated public in general. The writer of a college text-book in mathematics was naturally regarded as a great mathematician, despite the circumstance that, in general, the book contained the sum of the author's knowledge of the subject treated, much more than the average teacher's knowledge, and quite as much as the most capable youth was expected to master under the most favoring conditions. In general, neither author nor teacher nor pupil had knowledge of the fact that their most advanced instruction dealt only with the rudiments and often even with these in an obsolete or obsolescent manner; in general, there was no suspicion that, on the other side of the Atlantic, mathematics was a vast and growing science, much less that it was developing so rapidly and in so manifold a manner that the greatest mathematical genius found it necessary to specialize, even in his own domain. As a natural consequence American mathematical instruction depended almost exclusively on the use of text-books. What was thus at first a necessity became a tradition, and, accordingly, in striking contrast with French and German practice in schools of corre-sponding grade, American college and undergraduate uni-versity instruction in mathematics, with some exceptions, of which Harvard is the most notable, continues still to make the text-book the basis of instruction, even where it is not regarded as a *sine qua non* of the classroom. One result of this practice and tradition is that the text-book, which early assumed in the public estimation what now seems to be an exaggerated importance, continues still to be often regarded as an indispensable instrument for the systematic impartation of knowledge.

The text-book method in undergraduate mathematical

instruction undoubtedly has some peculiar merits and is recommended by considerations of weight. I am not about to advocate its abandonment. That question, moreover, is in a sense alien to the subject here under discussion. But I may say in passing that the notion, so firmly lodged in many of our colleges and still more firmly established in the mind of the general educated public, that the text-book is indispensable, is an erroneous one. That, as already said, has been amply proved both here and abroad, at Harvard, in some American normal schools, and in the schools of Germany and France, by the best, if not the only, method available for settling such questions, namely, by trial. And I could wish it were better known, particularly to teachers in secondary schools, that some of the ablest mathematicians and teachers of mathematics deprecate, not the use of the text-book, for that use has been sufficiently justified by the practice of most eminent and effective teachers, but our traditional dependence upon it, believing that this dependence often hampers the competent teacher's freedom and so prevents a full manifestation of the proper life of the subject. For the subject has indeed a deep and serene and even a joyous life, and, contrary to popular feeling, it is capable of being so interpreted and administered as to have, not merely for the few, but for the many, for the majority indeed of those who find their way to college, not only the highest disciplinary value, which is generally conceded, but a wonderful quickening power and inspiration as well. And it may very well be that the very great, though not generally suspected, human significance and cultural value of mathematics, the fact that not merely in its elements it is highly useful and applicable, but that throughout the entire immensity and wondrous complex of its development it is informed with beauty, being sustained indeed by artistic interest, — it may indeed be that all this will in some larger measure come to be felt and understood when teaching shall depend less on the text-book, at best a relatively dead thing, tending to bear the spirit of instruction down, and shall instead be more by living men, speaking immediately to living men, out of masterful knowledge of their science and with a clear perception of its spiritual significance and worth.

To return from this digression, it is fully recognized by all that, as undergraduate mathematical instruction is now carried on in our country, the text-book writer is a pretty valuable citizen, nor is there any disposition to detract from the dignity of his activity. Indeed, though some of the older books compare favorably in important respects with the best of the new, it may be said that, in general, to write a highly acceptable mathe-matical book for college use requires to-day an order of attainment far superior to that which was sufficient even a score or two of years ago. The training and scholar-ship which such work presupposes are, in respect to amount and more especially in respect to quality, not only relatively great, but very considerable absolutely. The author of the kind of book in question — and hap-pily there is no lack of competition in this field of writing — may be certain of intelligent, if not always generous, appreciation; he may be able thereby to lengthen his purse, his book stands some chance of being briefly noticed in reputable journals, and he may even gain local fame, but, however excellent the quality of his work-manship, it will seldom secure him a place in the ranks of the investigator or producer. The service of the text-book writer has not been degraded. It receives a more discriminating appreciation than ever before. It is merely that this kind of work has received a more critical appraisement. Not a few mathematicians decline to undertake the work of text-book writing, for the reason that they do not wish to be classed as text-book authors. If one who has published several original papers yields to the temptation to write a book for college use, the chances are his reputation will suffer loss rather than gain. Possibly such ought not to be the case, but nevertheless it is the case.

In regard to mathematical productivity proper, it is probably true that during the last twenty-five years, especially during the latter half of this period, there has been greater improvement in research work and output in this country than elsewhere in the world. Such sweeping statements are of course hazardous, and I make this one subject to correction. At all events, the gain in question has been great and is full of promise. Just about twenty-five years ago the *American Journal of Mathematics* was founded at the Johns Hopkins University, where it is still published as a quarterly. Previous to that time two or three attempts had been made to publish journals of mathematics in this country, but they met with little success, and are now scarcely remembered. The *American Journal of Mathematics*, in the beginning, sought contributions from abroad, and reference to the early volumes will show that these are to a considerable extent occupied by foreign products. A second journal, the *Annals of Mathematics*, was founded in 1884, and published at the University of Virginia. This journal, a quarterly, still flourishes, being now published at Cambridge, Mass., under the auspices of Harvard University. In 1888 was founded the New York Mathematical Club, which soon became the New York Mathematical Society and began the publication of a monthly *Bulletin*. In

1894 this society became the American Mathematical Society which now has a membership of nearly four hundred, including, with few exceptions, every American mathematician of standing, besides some members from Canada, England, and the Continent. This society has a rapidly growing library, and publishes two journals, the *Bulletin,* already mentioned, and the *Transactions,* a quarterly journal, recently founded, and devoted to the publication of the more important results of research. The four journals named are, all of them, of good standing and exchange with some of the best British and Continental journals. Not by any means all the members of the society are producing mathematicians, but a large percentage of them are sufficiently interested to attend one or more meetings of the society each year. These meetings are bi-monthly meetings, held in New York, and a summer meeting at a place chosen from year to year. To meet growing demands, a Chicago section has been organized, which holds regular meetings in that city, and a second section on the Pacific Coast, whose business will be conducted perhaps at San Francisco. These sections report to the society proper, which has its offices in New York City.

At these meetings there are presented annually several scores of papers, a percentage of which deal with applied mathematics. Of course not all of these papers are important, but some of them possess very considerable, a few of them distinctly great, value, and a large majority of them fall properly within the category of original investigation as defined. In addition to such more regular contributions, a considerable number of mathematical papers are annually presented before other American scientific organizations, as, for example, before Section A of the American Association for the Advancement of

Science. The majority of all these articles are found to be available for publication, and the result is that, although foreign contributions are no longer invited as formerly, and few of them received, the four journals above mentioned are, nevertheless, taxed beyond their capacity; and, for want of room, papers are sometimes rejected by the American journals which, if produced abroad, would probably be published there, where the facilities for publication are ampler. In fact, the number of American memoirs published abroad exceeds perhaps the number of foreign contributions published here.

It is greatly to be regretted that our facilities for publication, though recently so greatly enhanced, are still distinctly inadequate. For mathematicians are also men, and, as such, one of their most powerful incentives to research is the prospect of the recognition that comes from having the results of their labors properly placed before the scientific world.

While the picture thus drawn of American mathematical activity is a pleasing one and is full of encouragement and hope, still we must not disguise from ourselves the fact that, in view of the vast extent and resources of our country and of the large number of professional mathematicians connected with our numerous colleges and universities, the amount and the average quality of the American mathematical output are not only distinctly inferior to that of the more scientific countries of Europe, for which, not without some justice and plausibility, we are wont to plead our youth in defense and explanation, but this average and amount are by no means a measure of our native ability nor in keeping with our achievements in some other scarcely worthier, if less ethereal, domains.

The reasons for this state of case are not far to seek, and come readily to light on a minuter study of the necessary and sufficient conditions for the vigorous prosecution of mathematical research.

We may recall the philosopher's insight that " there is but one poet and that is Deity." The poet is indeed born, we all agree; and it is equally, if not so obviously, true that the great mathematician or financier or administrator is born. But the mathematician is not born trained or born with knowledge of the state of the science, and hence it goes without saying that to native ability, which we presuppose throughout as absolutely essential and which is not so rare as is often thought, training must be superadded, years of austere training under, or still better, in co-operation with, competent masters in a suitable atmosphere. Formerly, it was in general necessary to seek such training abroad; that is no longer the case, now that our better universities are manned with scholars of the best American and European training. Indeed the mathematical doctorate of a few of our own institutions now represents quite as much as, if not more than, the average German doctorate, though less, we must still confess, than the French, which probably has the highest significance of any in the world. Several of the most highly productive mathematicians in the country have not received foreign training, while a still larger number of non-producers studied abroad for years — a fact showing that such training is neither a necessary nor a sufficient condition. It is not intended to depreciate the absolute value of foreign training, but only its relative value — its value as compared with that of the best which our own country now affords. It is still desirable, when not too inconvenient, to spend a year in the atmosphere of foreign universities, and many avail themselves of the opportunity, largely for the sake of the prestige which, owing partly to a tradition, it still affords in many American communities. It is, of course, a mere truism to say that training, though necessary, is not sufficient. Unless there be the gift of originality, training can at best result in receptive and critical scholarship, but not in productive power.

There are in our country a goodly number of men having the requisite ability and training, who, nevertheless, produce but little or nothing at all — a fact to be accounted for by the absence in their case of other essential conditions.

In some cases library facilities are lacking. Mathematical science is a growth. The new rises out of the old, whence the necessity that the investigator have at hand the major part at least of the literature of his subject from the earliest times. Even more exacting, if possible, is the necessity of having ready access to the leading journals of England, Germany, France, and Italy, besides those of America. One takes special pleasure in mentioning Italy, because she has been recently making rapid advances and in two important directions, the geometry of hyperspace and mathematical logic, the ontology of pure thought, she comes well-nigh leading the van. Of journals there are at least a dozen which are absolutely indispensable to the research worker and as many more that are highly desirable. In addition, the producing mathematician will not infrequently have occasion to refer to memoirs which, because of their length or for other reasons, have not appeared in the journals, and are to be found only in the proceedings of the leading general scientific and philosophical societies of Europe. The lack of such facilities, which in some cases a few hundred and in others a few thousand dollars would suffice to make good, is in itself sufficient to explain the non-productivity of not a few American mathematicians.

Again, there are cases where able men have not the necessary leisure to engage successfully in investigation. Our universities are for the most part so organized that the energies of scientific men are largely expended in undergraduate teaching and in administrative work. We, as a people, have yet to learn that the value of a professor to a community can be rightly estimated, not by counting the number of hours he actually stands before his classes, but rather, if we must count at all, by reckoning the number of hours devoted to the preparation of his lectures, and more particularly by the fruit of quiet study and research. In Germany the ordinary professor lectures from four to six hours a week, to which if we add in some cases two hours *Seminarübungen,* we have a total of six to eight hours of presence in the lecture room. In France the professor is expected to give one course of lectures. These take place twice a week and last from one to one and a half hours. To this duty should be added that of holding a large number of examinations — a rather wearisome service from which the German escapes. When we contrast this with the ten to fifteen and often even twenty or more hours of actual teaching demanded of the American professor, to say nothing of faculty meetings, committee meetings, and the multitudinous examinations, and when we do not fail to reflect that ten hours are much more than twice five in their tax upon energy, it is little wonder that in productivity our most brilliant men are often so greatly outclassed by their foreign competitors. Moreover, " our universities are at present, for the first two years, gymnasia and lycées, and our professors are accordingly obliged to devote themselves largely to what is properly secondary instruction " — a kind of work which, however worthy, important, and necessary, has the effect, not merely of drawing off the energy in non-productive channels, but also eventually of forming and hardening the mind about a relatively small group of simpler ideas.

Again, scientific activity is not infrequently rendered impossible by the amount of administrative work which professors of notable administrative ability are called upon to perform. Indeed " the problem presses for solution, how to retain the many peculiar excellences of our college and university life and at the same time to create for certain men of talent and training a suitable environment for the highest scientific activity."

Once more, it is very desirable, indeed it is really necessary, for men working in a branch of science to attend the meetings of scientific bodies, in order to meet their fellow-men, to take counsel of them, to create and share in a wholesome *esprit de corps*, to catch the inspiration and enthusiasm, and to gain the sustaining impulses which can come only from personal contact and co-operation. But our country is so vast, the distances so long, and traveling so expensive, that many mathematicians, owing to smallness of income, find themselves hopelessly condemned to a life of isolation, of which the result is a loss first of interest and then of power. It is in vain that one counsels such men to wake up and be strong and active, for their state of inactivity is less a defect of will than an effect of circumstances.

There is a second phase of this question of remuneration which is, happily, beginning to attract attention and to receive consideration in university circles. I refer to the proposition that a wise economy will provide, for university service, remuneration, not such as would attract men whose first ambition is to acquire the ease that wealth is supposed to afford, but such as will not, by its inadequacy to the reasonable demands of modern social life, deter men of ability and predilection for scientific pursuits from entering upon them. I know personally of six young men, not all of them mathematicians, who have sufficiently demonstrated that they possess such ability and predilection, five of whom have recently relinquished the pursuit of science and the fifth of whom told me only yesterday that he seriously contemplates doing so, all of them, for the reason that, as they allege, the university career furnishes either not at all, or too tardily, a financial competence and consequent relief from practical condemnation to celibacy. It matters little whether they be mistaken to a degree or not, so long as the contrary conviction determines choice. There is, indeed, more than a bare suspicion that for reasons akin to those actuating in the cases cited, the university career, particularly in case of the more abstract sciences, such as pure mathematics, whose doctrines have little or no market value, fails to attract a due proportion of the best intellects of the country. For it should be understood that successful investigation in such sciences demands men of intellectual resource, of power, of persistence, in a word, men of strenuosity of life and character. Such men are indeed the intellectual peers of the great financier, or soldier, or statesman, or administrator, and they are aware of it; so that if too many such men are not to be drawn away from scientific fields by the prospect of achieving elsewhere not only fame but fortune also, it stands to reason that the university career must promise at least a competence and the peace of mind it brings. That such is the case, and that the future will condemn the present for a too tardy recognition of the fact, is a matter which can hardly admit of doubt.

That the conditions above indicated are those which

determine the matter of mathematical productivity is a proposition which not only commends itself *a priori* to the reason, but is justified also *a posteriori* by experience, for statistics show that those institutions, both foreign and domestic, where such productivity has flourished best are also those where the conditions named are most fully satisfied, and that where one, at least, of the conditions is not fulfilled, there investigation proceeds but feebly or is wholly wanting.

It remains to mention another condition which in a sense includes all others, and whose fulfillment will come gradually as at once the cause and the effect of the fulfillment of all others. I mean, of course, a public sentiment which will demand, because it has learned to appreciate, knowledge, not merely because of its applications and utility, but for its beauty, as one appreciates the moon and the stars without regard to their aid in navigation — a public sentiment that shall seek every provision and regard as sacred every instrumentality for the advancement and ministration of knowledge, not only as a means to happiness, but as a glory, for its own sake, as a self-justifying realization of the distinctive ambition of man, to understand the universe in which he lives and the wondrous possibilities of the Reason unto which it constantly makes appeal.

MATHEMATICS [1]

IN the early part of the last century a philosophic
French mathematician, addressing himself to the question of the perfectibility of scientific doctrines, expressed
the opinion that one may not imagine the last word has
been said of a given theory so long as it can not by
a brief explanation be made clear to the man of the
street. Doubtless that conception of doctrinal perfectibility, taken literally, can never be realized. For doubtless, just as there exist now, so in the future there
will abound, even in greater and greater variety and on
a vaster and vaster scale, deep-laid and high-towering
scientific doctrines that, in respect to their infinitude of
detail and in their remoter parts and more recondite
structure, shall not be intelligible to any but such as
concentrate their life upon them. And so the noble dream
of Gergonne can never literally come true. Nevertheless,
as an ideal, as a goal of aspiration, it is of the highest
value, and, though in no case can it be quite attained, it
yet admits in many, as I believe, of a surprisingly high
degree of approximation. I do not mind frankly owning
that I do not share in the feeling of those, if there be

[1] An address delivered in 1907 at Columbia University, the University of Virginia, Washington and Lee University, the University of
North Carolina, Tulane University, the University of Arkansas, the
University of Nebraska, the University of Missouri, the University of
Chicago, Northwestern University, the University of Illinois, Vanderbilt
University, the University of Minnesota, the University of Michigan,
the University of Cincinnati, the Ohio State University, Vassar College,
the University of Vermont, Purdue University, and the University Club
of New York City. Printed by the Columbia University Press, 1907.

any such, who regard their special subjects as so intricate, mysterious and high, that in all their sublimer parts they are absolutely inaccessible to the profane man of merely general culture even when he is led by the hand of an expert and condescending guide. For scientific theories are, each and all of them, and they will continue to be, built upon and about notions which, however sublimated, are nevertheless derived from common sense. These etherealized central concepts, together with their manifold bearings on the higher interests of life and general thought, can be measurably assimilated to the language of the common level from which they arose. And, in passing, I should like to express the hope that here at Columbia or other competent center there may one day be established a magazine that shall have for its aim to mediate, by the help, if it may be found, of such pens as those of Huxley and Clifford, between the focal concepts and the larger aspects of the technical doctrines of the specialist, on the one hand, and the teeming curiosity, the great listening, waiting, eager, hungering consciousness of the educated thinking public on the other. Such a service, however, is not to be lightly undertaken. An hour, at all events, is hardly time enough in which to conduct an excursion even of scientific folk through the mazes of more than twenty hundred years of mathematical thought or even to express intelligibly, if one were competent, the significance of the whole in a critical estimate.

Indeed, such is the character of mathematics in its profounder depths and in its higher and remoter zones that it is well nigh impossible to convey to one who has not devoted years to its exploration a just impression of the scope and magnitude of the existing body of the science. An imagination formed by other disciplines and accustomed to the interests of another field may scarcely receive suddenly an apocalyptic vision of that infinite interior world. But how amazing and how edifying were such a revelation, if only it could be made. To tell the story of mathematics from Pythagoras and Plato to Hilbert and Lie and Poincaré; to recount and appraise the achievements of such as Euclid and Archimedes, Apollonius and Diophantus; to display and estimate the creations of Descartes and Leibniz and Newton; to dispose in genetic order, to analyze, to synthesize and evaluate, the discoveries of the Bernoullis and Euler, of Desargues and Pascal and Monge and Poncelet, of Steiner and Möbius and Plücker and Staudt, of Lobatschewsky and Bolyai, of W. R. Hamilton and Grassmann, of Laplace, Lagrange and Gauss, of Boole and Cayley and Hermite and Gordan, of Bolzano and Cauchy, of Riemann and Weierstrass, of Georg Cantor and Boltzmann and Klein, of the Peirces and Schroeder and Peano, of Helmholtz and Maxwell and Gibbs; to explore, and then to map for perspective beholding and contemplation, the continent of doctrine built up by these immortals, to say nothing of the countless refinements, extensions and elaborations meanwhile wrought by the genius and industry of a thousand other agents of the mathetic spirit; — to do that would indeed be to render an exceeding service to the higher intelligence of the world, but a service that would require the conjoint labors of a council of scholars for the space of many years. Even the three immense volumes of Moritz Cantor's *Geschichte der Mathematik,* though they do not aspire to the higher forms of elaborate exposition and though they are far from exhausting the material of the period traversed by them, yet conduct the narrative down only to 1758.[2]

[2] The work is now being carried forward by younger men.

That date, however, but marks the time when mathematics, then schooled for over a hundred eventful years in the unfolding wonders of Analytic Geometry and the Calculus and rejoicing in the possession of these the two most powerful among the instruments of human thought, had but fairly entered upon her modern career. And so fruitful have been the intervening years, so swift the march along the myriad tracks of modern analysis and geometry, so abounding and bold and fertile withal has been the creative genius of the time, that to record even briefly the discoveries and the creations since the closing date of Cantor's work would require an addition to his great volumes of a score of volumes more.

Indeed the modern developments of mathematics constitute not only one of the most impressive, but one of the most characteristic, phenomena of our age. It is a phenomenon, however, of which the boasted intelligence of a "universalized" daily press seems strangely unaware; and there is no other great human interest, whether of science or of art, regarding which the mind of the educated public is permitted to hold so many fallacious opinions and inferior estimates. The golden age of mathematics — that was not the age of Euclid, it is ours. Ours is the age in which no less than six international congresses of mathematics have been held in the course of nine years.[3] It is in our day that more than a dozen mathematical societies contain a growing membership of over two thousand men representing the centers of scientific light throughout the great culture nations of the world. It is in our time that over five hundred scientific journals are each devoted in part,

[3] International congresses of mathematicians are held at intervals of four years. Since the date of this address two have been held, one at Rome in 1908 and one in Cambridge, England, in 1912.

while more than two score others are devoted exclusively, to the publication of mathematics. It is in our time that the *Jahrbuch über die Fortschritte der Mathematik,* though admitting only condensed abstracts with titles, and not reporting on all the journals, has, nevertheless, grown to nearly forty huge volumes in as many years. It is in our time that as many as two thousand books and memoirs drop from the mathematical press of the world in a single year, the estimated number mounting up to fifty thousand in the last generation. Finally, to adduce yet another evidence of similar kind, it requires no less than the seven ponderous tomes of the forthcoming *Ency-klopädie der Mathematischen Wissenschaften* to contain, not expositions, not demonstrations, but merely compact reports and bibliographic notices sketching developments that have taken place since the beginning of the nine-teenth century. The Elements of Euclid is as small a part of mathematics as the Iliad is of literature; or as the sculpture of Phidias is of the world's total art. Indeed if Euclid or even Descartes were to return to the abode of living men and repair to a university to resume pursuit of his favorite study, it is evident that, making due allow-ance for his genius and his fame, and presupposing familiarity with the modern scientific languages, he would yet be required to devote at least a year to preparation before being qualified even to begin a single strictly graduate course.

It is not, however, by such comparisons nor by sta-tistical methods nor by any external sign whatever, but only by continued dwelling within the subtle radiance of the discipline itself, that one at length may catch the spirit and learn to estimate the abounding life of modern mathesis: oldest of the sciences, yet flourishing to-day as never before, not merely as a giant tree throwing out and aloft myriad branching arms in the upper regions of clearer light and plunging deeper and deeper root in the darker soil beneath, but rather as an immense mighty forest of such oaks, which, however, literally grow into each other so that by the junction and intercrescence of limb with limb and root with root and trunk with trunk the manifold wood becomes a single living organic growing whole.

What is this thing so marvelously vital? What does it undertake? What is its motive? What its significance? How is it related to other modes and forms and interests of the human spirit?

What is mathematics? [4] I inquire, not about the word, but about the thing. Many have been the answers of former years, but none has approved itself as final. All of them, by nature belonging to the " literature of knowledge," have fallen under its law and " perished by supersession." Naturally conception of the science has had to grow with the growth of the science itself.

A traditional conception, still current everywhere except in critical circles, has held mathematics to be the science of quantity or magnitude, where magnitude including multitude (with its correlate of number) as a special kind, signified whatever was " capable of increase and decrease and measurement." Measurability was the essential thing. That definition of the science was a very natural one, for magnitude did appear to be a singularly fundamental notion, not only inviting but demanding consideration at every stage and turn of life. The necessity of finding out how many and how much was the mother of counting and measurement, and mathematics, first from necessity and then from pure curiosity and joy,

---

[4] For a deeper discussion see Keyser's *Mathematical Philosophy* (Dutton & Company).

so occupied itself with these things that they came to seem its whole employment.

Nevertheless, numerous great events of a hundred years have been absolutely decisive against that view. For one thing, the notion of the *continuum* — the " Grand Continuum " as Sylvester called it — that great central supporting pillar of modern Analysis, has been constructed by Weierstrass, Dedekind, Georg Cantor and others, without any reference whatever to quantity, so that number and magnitude are not only independent, they are essentially disparate. When we attempt to correlate the two, the ordinary concept of measurement as the repeated application of a constant finite unit, undergoes such refinement and generalization through the notion of Limit [5] or its equivalent that counting no longer avails and measurement retains scarcely a vestige of its original meaning. And when we add the further consideration that non-Euclidean geometry employs a scale in which the unit of angle and distance, though it is a constant unit, nevertheless appears from the Euclidean point of view to suffer lawful change from step to step of its application, it is seen that to retain the old words and call mathematics the science of quantity or magnitude, and measurement, is quite inept as no longer telling either what the science has actually become or what its spirit is bent upon.

Moreover, the most striking measurements, as of the volume of a planet, the growth of cells, the valency of atoms, rates of chemical change, the swiftness of thought, the penetrative power of radium emanations, are none of them done by *direct* repeated application of a unit or by any direct method whatever. They are all of them

[5] For an evaluation of this sovereign concept see the treatment of it in Keyser's *Mathematical Philosophy*.

accomplished by one form or another of indirection. It was perception of this fact that led the famous philosopher and respectable mathematician, Auguste Comte, to define mathematics as " the science of indirect measurement." Here doubtless we are in presence of a finer insight and a larger view, but the thought is not yet either wide enough or deep enough. For it is obvious that there is an immense deal of admittedly mathematical activity that is not in the least concerned with measurement whether direct or indirect. Consider, for example, that splendid creation of the nineteenth century known as Projective Geometry: a boundless domain of countless fields where reals and imaginaries, finites and infinites, enter on equal terms, where the spirit delights in the artistic balance and symmetric interplay of a kind of conceptual and logical counterpoint, — an enchanted realm where thought is double and flows throughout in parallel streams. Here there is no essential concern with number or quantity or magnitude, and metric considerations are entirely absent or completely subordinate. The fact, to take a simplest example, that two points determine a line uniquely, or that the intersection of a sphere and a plane is a circle, or that any configuration whatever — the reference is here to ordinary space — presents two reciprocal aspects according as it is viewed as an ensemble of points or as a manifold of planes, is not a *metric* fact at all: it is not a fact about size or quantity or magnitude of any kind. In this domain it was *position* rather than size that seemed to some the central matter, and so it was proposed to call mathematics the science of measurement and *position*.

Even as thus expanded, the conception yet excludes many a mathematical realm of vast extent. Consider that immense class of things known as Operations. These are limitless alike in number and in kind. Now it so happens that there are many systems of operations such that any two operations of a given system, if thought as following one another, together thus produce the same effect as some other single operation of the system. Such systems are infinitely numerous and present themselves on every hand. For a simple illustration, think of the totality of possible straight motions in space. The operation of going from point A to point B, followed by the operation of going from B to point C, is equivalent to the single operation of going straight from A to C. Thus the system of such operations is a closed system: combination, *i.e.*, of any two of the operations yields a third one, not without, but within, the system. The great notion of Group, thus simply exemplified, though it had barely emerged into consciousness a hundred years ago, has meanwhile become a concept of fundamental importance and prodigious fertility, not only affording the basis of an imposing doctrine — the Theory of Groups — but therewith serving also as a bond of union, a kind of connective tissue, or rather as an immense cerebro-spinal system, uniting together a large number of widely dissimilar doctrines as organs of a single body. But — and this is the point to be noted here — the abstract operations of a group, though they are very real things, are neither magnitudes nor positions.

This way of trying to come to an adequate conception of mathematics, namely, by attempting to characterize in succession its distinct domains, or its varieties of content, or its modes of activity, in the hope of finding a common definitive mark, is not likely to prove successful. For it demands an exhaustive enumeration, not only of the fields now occupied by the science, but also of those destined to be conquered by it in the future, and such an achievement would require a prevision that none may claim.

Fortunately there are other paths of approach that seem more promising. Everyone has observed that mathematics, whatever it may be, possesses a certain mark, namely, a degree of certainty not found elsewhere. So it is, proverbially, the exact science par excellence. Exact, no doubt, but in what sense? An excellent answer is found in a definition given about one generation ago by a distinguished American mathematician, Professor Benjamin Peirce: " Mathematics is the science which draws necessary conclusions " — a formulation of like significance with the following fine *mot* by Professor William Benjamin Smith: " Mathematics is the universal art apodictic." These statements, though neither of them is adequate, are both of them telling approximations, at once foreshadowing and neatly summarizing for popular use, the epoch-making thesis established by the creators of modern logic, namely, that mathematics is included in, and, in a profound sense, may be said to be identical with, Symbolic Logic. Observe that the emphasis falls on the quality of being " necessary," *i.e.*, correct logically, or valid formally.

But why are mathematical conclusions correct? Is it that the mathematician has a reasoning faculty essentially different in kind from that of other men? By no means. What, then, is the secret? Reflect that conclusion implies premises, that premises involve terms, that terms stand for ideas or concepts or notions, and that these latter are the ultimate material with which the spiritual architect, called the Reason, designs and builds. Here, then, one may expect to find light. The apodictic quality of mathematical thought, the correctness of its conclusions as conclusions, are due, not to any special mode of ratiocination, but to the character of the concepts with which it deals. What is that distinctive characteristic? The answer is: precision and completeness of determination. But how comes the mathematician by such completeness? There is no mysterious trick involved: some concepts admit of such precision and completeness, others do not; the mathematician is one who deals with those that do.

The matter, however, is not quite so simple as it sounds, and I bespeak your attention to a word of caution and of further explanation. The ancient maxim, *ex nihilo nihil fit*, may well be doubted where it seems most obviously valid, namely, in the realm of matter, for it may be that matter has evolved from something else; but the maxim cannot be ultimately denied where its application is least obvious, namely, in the realm of mind, for without principia in the strictest sense, doctrine is, in the strictest sense, impossible. And when the mathematician speaks of complete determination of concepts and of rigor of demonstration, he does not mean that the undefined and the undemonstrated have been or can be entirely eliminated from the foundations of his science. He knows that such elimination is impossible; he knows, too, that it is unnecessary, for some undefinable ideas are perfectly clear and some undemonstrable propositions are perfectly precise and certain. It is in terms of such concepts that a definable notion, if it is to be mathematically available, must admit of complete determination, and in terms of such propositions that mathematical discourse secures its rigor. It is, then, of such indefinables among ideas and such indemonstrables among propositions — paradoxical as the statement may appear — that the foundations of mathematics in its ideal conception are composed; and whatever doctrine is logically constructible on such a basis is mathematics either actually or potentially. I am not asserting that the substructure herewith characterized has been brought to completion. It is on the conception of it that the accent is here designed to fall, for it is the conception as such that at once affords to fundamental investigation a goal and a guide and furnishes the means of giving the science an adequate definition.

On the other hand, actually to realize the conception requires that the foundation to be established shall both include every element that is essential and exclude every one that is not. For a foundation that subsequently demands or allows superfoetation of hypotheses is incomplete; and one that contains the non-essential is imperfect. Of the two problems thus presented, it is the latter, the problem of exclusion, of reducing principles to a minimum, of applying Occam's Razor to the pruning away of non-essentials, — it is that problem that taxes most severely both the analytic and the constructive powers of criticism. And it is to the soluton of that problem that the same critical spirit of our time, which in other fields is reconstructing theology, burning out the dross from philosophy, and working relentless transformations of thought on every hand, has directed a chief movement of modern mathematics.

Apart from its technical importance, which can scarcely be overestimated, the power, depth and comprehensiveness of the modern critical movement in mathematics, make it one of the most significant scientific phenomena of the last century. Double in respect to origin, the movement itself has been composite. One component began at the very center of mathematical activity, while the other took its rise in what was then erroneously regarded as an alien domain, the great domain of symbolic logic.

A word as to the former component. For more than a hundred years after the inventions of Analytical Geometry and the Calculus, mathematicians may be said to have fairly rioted in applications of these instruments to physical, mechanical and geometric problems, without concerning themselves about the nicer questions of fundamental principles, cogency, and precision. In the latter part of the eighteenth century the efforts of Euler, Lacroix and others to systematize results served to reveal in a startling way the necessity of improving foundations. Constructive work was not indeed arrested by that disclosure. On the contrary new doctrines continued to rise and old ones to expand and flourish. But a new spirit had begun to manifest itself. The science became increasingly critical as its towering edifices more and more challenged attention to their foundations. Manifest already in the work of Gauss and Lagrange, the new tendence, under the powerful impulse and leadership of Cauchy, rapidly develops into a momentous movement. The Calculus, while its instrumental efficacy is meanwhile marvelously improved, is itself advanced from the level of a tool to the rank and dignity of a science. The doctrines of the real and of the complex variable are grounded with infinite patience and care, so that, owing chiefly to the critical constructive genius of Weierstrass and his school, that stateliest of all the pure creations of the human intellect — the Modern Theory of Functions with its manifold branches — rests to-day on a basis not less certain and not less enduring than the very integers with which we count. The movement still sweeps on, not only extending to all the cardinal divisions of Analysis but, through the agencies of such as Lobatschewsky and Bolyai, Grassmann and Riemann, Cayley and Klein, Hilbert and Lie, recasting the foundations of Geometry also. And there can scarcely be a doubt that the great domains of Mechanics and Mathematical Physics are by their need destined to a like invasion.

In the light of all this criticism, mathematics came to appear as a great ensemble of theories, compendent no doubt, interpenetrating each other in a wondrous way, yet all of them distinct, each built up by logical processes on its own appropriate basis of pure hypotheses, or assumptions, or postulates. As all the theories were thus seen to rest equally on hypothetical foundations, all were seen to be equally legitimate; and doctrines like those of Quaternions, non-Euclidean geometry and Hyperspace, for a time suspected because based on postulates not all of them traditional, speedily overcame their heretical reputations and were admitted to the circle of the lawful and orthodox.

It is one thing, however, to deal with the principal divisions of mathematics severally, underpinning each with a foundation of its own. That, broadly speaking, has been the plan and the effect of the critical movement as thus far sketched. But it is a very different and a profounder thing to underlay all the divisions at once with a single foundation, with a foundation that shall serve as a support, not merely for all the *divisions* but for something else, distinct from each and from the sum of all, namely, for the *whole,* the science itself, which they constitute. It is nothing less than that achievement which, unconsciously at first, consciously at last, has been the aim and goal of the other component of the critical movement, that component which, as already said, found its origin and its initial interest in the field of symbolic logic. The advantage of employing symbols in the investigation and exposition of the formal laws of thought is not a recent discovery. As everyone knows, symbols were thus employed to a small extent by the Stagirite himself. The advantage, however, was not pursued; because for two thousand years the eyes of logicians were blinded by the blazing genius of the "master of those that know." With the single exception of the reign of Euclid, the annals of science afford no match for the tyranny that has been exercised by the logic of Aristotle. Even the important logical researches of Leibniz and Lambert and their daring use of symbolic methods were powerless to break the spell. It was not till 1854 when George Boole, having invented an algebra to trace and illuminate the subtle ways of reason, published his symbolical "Investigation of the Laws of Thought," that the revolution in logic really began. For, although for a time neglected by logicians and mathematicians alike, it was Boole's work that inspired and inaugurated the scientific movement now known and honored throughout the world under the name of Symbolic Logic.

It is true, the revolution has advanced in silence. The discoveries and creations of Boole's successors, of C. S. Peirce, of Schroeder, of Peano and of their disciples and peers, have not been proclaimed by the daily press. Commerce and politics, gossip and sport, accident and crime, the shallow and transitory affairs of the exoteric world, — these have filled the columns and left no room to publish abroad the deep and abiding things achieved in the silence of cloistral thought. The demonstration by symbolical means of the fact that the three laws of Identity, Excluded Middle and Non-contradiction are absolutely independent, none of them being derivable from the other two; the discovery that the syllogism is not deducible from those laws but has to be postulated as an independent principle; the discovery of the astounding and significant fact that false propositions imply all propositions and that true ones, though not implying, are implied by, all; the discovery that most reasoning is not syllogistic, but is asyllogistic, in form, and that, therefore, contrary to the teaching of tradition, the class-logic of Aristotle is not adequate to all the concerns of rigorous thought; the discovery that Relations, no less than Classes, demand a logic of their own, and that a similar claim is valid in the case of Propositions: no intelligence of these events nor of the immense multitude of others which they but meagrely serve to hint and to exemplify, has been cabled round the world and spread broadcast by the flying bulletins of news. Even the scientific public, for the most part accustomed to viewing the mind as only the instrument and not as a subject of study, has been slow to recognize the achievements of modern research in the minute anatomy of thought. Indeed it has been not uncommon for students of natural science to sneer at logic as a stale ad profitless pursuit, as the barren mistress of scholastic minds.[6] These men have not been aware of what certainly is a most profound, if indeed it be not the most significant, scientific movement of our time. In America, in England, in Germany, in France, and especially in Italy — supreme histologist of the human understanding — the deeps of mind and logical reality have been explored in our generation as never before in the history of the world. Owing to the power of the symbolic method, not only the foundations of the Aristotelian logic — the Calculus of Classes — have been recast, but side by side with that everlasting monument of Greek genius, there rise today other structures, fit companions of the ancient edifice, namely, the Logic of Relations and the Logic of Propositions.

[6] The infinite importance of logic for human welfare is impressively set forth for the general reader in Polakov's *Man and His Affairs,* The Williams and Wilkins Company.

And what are the entities that have been found to constitute the base of that triune organon? The answer is surprising: a score or so of primitive, indemonstrable, propositions together with less than a dozen undefinable notions, called logical constants. But what is more surprising — for here we touch the goal and are enabled to enunciate what has been justly called " one of the greatest discoveries of our age " — is the fact that the basis of logic is the basis of mathematics also. Thus the two great components of the critical movement, though distinct in origin and following separate paths, are found to converge at last in the thesis: Symbolic Logic is Mathematics, Mathematics is Symbolic Logic, the twain are one.

Is it really so? Does the identity exist in fact? Is it true that so simple a unifying foundation for what has hitherto been supposed two distinct and even mutually alien interests has been actually ascertained? The basal masonry is indeed not yet completed but the work has advanced so far that the thesis stated is beyond dispute or reasonable doubt. Primitive propositions appear to allow some freedom of choice, questions still exist regarding relative fundamentality, and statements of principles have not yet crystallized into settled and final form; but regarding the nature of the data to be assumed, the smallness of their number and their adequacy, agreement is substantial. In England, Russell and Whitehead [7] are successfully engaged now in forging " chains of deduction " binding the cardinal matters of Analysis and Geometry to the premises of General Logic, while in Italy the *Formulaire de Mathématiques* of Peano and his school has been for some years growing into a veritable

[7] This work has been projected in four immense volumes bearing the title, *Principia Mathematica*, of which three volumes have appeared.

encyclopedia of mathematics wrought by the means and clad in the garb of symbolic logic.

But is it not incredible that the concept of number with all its distinctions of cardinal and ordinal, fractional and whole, rational and irrational, algebraic and transcendental, real and complex, finite and infinite, and the concept of geometric space, in all its varieties of form and dimensionality, is it not incredible that mathematical ideas, surpassing in multitude the sands of the sea, should be precisely definable, each and all of them, in terms of a few logical constants, in terms, *i.e.*, of such indefinable notions as *such that, implication, denoting, relation, class, propositional function,* and two or three others? And is it not incredible that by means of so few as a score of premises (composed of ten principles of deduction and ten other indemonstrable propositions of a general logical nature), the entire body of mathematical doctrine can be strictly and formally deduced?

It is wonderful, indeed, but not incredible. Not incredible in a world where the mustard seed becometh a tree, not incredible in a world where all the tints and hues of sea and land and sky are derived from three primary colors, where the harmonies and the melodies of music proceed from notes that are all of them but so many specifications of four generic marks, and where three concepts — energy, mass, motion, or mass, time, space — apparently suffice for grasping together in organic unity the mechanical phenomena of a universe.

But the thesis granted, does it not but serve to justify the cardinal contentions of the depreciators of mathematics? Does it not follow from it that the science is only a logical grind, suited only to narrow and straitened intellects content to tramp in treadmill fashion the weary rounds of deduction? Does it not follow that Schopenhauer was right in regarding mathematics as the lowest form of mental activity, and that he and our own genial and enlightened countryman, Oliver Wendell Holmes, were right in likening mathematical thought to the operations of a calculating machine? Does it not follow that Huxley's characterization of mathematics as " that study which knows nothing of observation, nothing of induction, nothing of experiment, nothing of causation," is surprisingly confirmed by fact? Does it not follow that Sir William Hamilton's famous and terrific diatribe against the science finds ample warrant in truth? Does it not follow, as the Scotch philosopher maintains, that mathematics regarded as a discipline, as a builder of mind, is inferior? That devotion to it is fatal to the development of the sensibilities and the imagination? That continued pursuit of the study leaves the mind narrow and dry, meagre and lean, disqualifying it both for practical affairs and for those large and liberal studies where moral questions intervene and judgment depends, not on nice calculation by rule, but on a wide survey and a balancing of probabilities?

The answer is, No. Those things not only do not follow but they are not true. Every count in the indictment, whether explicit or only implied, is false. Not only that, but the opposite in each case is true. On that point there can be no doubt; authority, reason and fact, history and theory, are here in perfect accord. Let me say once for all that I am conscious of no desire to exaggerate the virtues of mathematics. I am willing to admit that mathematicians do constitute an important part of the salt of the earth. But the science is no catholicon for mental disease. There is in it no power for transforming mediocrity into genius. It cannot enrich where nature has impoverished. It makes no pretense of creat-

ing faculty where none exists, of opening springs in desert
minds. *" Du bist am Ende — was du bist."* The great
mathematician, like the great poet or great naturalist or
great administrator, is born. My contention shall be
that where the mathetic endowment is found, there will
usually be found associated with it, as essential implications in it, other endowments in generous measure, and
that the appeal of the science is to the whole mind,
direct no doubt to the central powers of thought, but
indirectly through sympathy of all, rousing, enlarging,
developing, emancipating all, so that the faculties of will,
of intellect and feeling, learn to respond, each in its
appropriate order and degree, like the parts of an or-
chestra to the " urge and ardor " of its leader and lord.

As for Hamilton and Schopenhauer, those detractors
need not detain us long. Indeed but for their fame and
the great influence their opinions have exercised over
" the ignorant mass of educated men," they ought not
in this connection to be noticed at all. Of the subject
on which they presumed to pronounce authoritative judgment of condemnation, they were both of them ignorant,
the former well nigh proudly so, the latter unawares, but
both of them, in view of their pretensions, disgracefully
ignorant. Lack of knowledge, however, is but a venial
sin, and English-speaking mathematicians have been disposed to hope that Hamilton might be saved in accordance with the good old catholic doctrine of invincible
ignorance. But even that hope, as we shall see, must be
relinquished. In 1853 William Whewell, then fellow and
tutor of Trinity College, Cambridge, published an appre-
ciative pamphlet entitled " Thoughts on the Study of
Mathematics as a Part of a Liberal Education." The
author was a brilliant scholar. " Science was his forte,"
but " omniscience his foible," and his reputation for

universal knowledge was looming large. That reputation, however, Hamilton regarded as his own prerogative. None might dispute the claim, much less share the glory of having it acknowledged on his own behalf. Whewell must be crushed. In the following year Sir William replies in the *Edinburg Review,* and such a show of learning! The reader is apparently confronted with the assembled opinions of the learned world, and — what is more amazing — they all agree. Literati of every kind, of all nations and every tongue, orators, philosophers, educators, scientific men, ancient and modern, known and unknown, all are made to support Hamilton's claim, and even the most celebrated mathematicians seem eager to declare that the study of mathematics is unworthy of genius and injures the mind. Whewell was overwhelmed, reduced to silence. His promised rejoinder failed to appear. The Scotchman's victory was complete, his fame enhanced, and his alleged judgment regarding a great human interest of which he was ignorant has reigned over the minds of thousands of men who have been either willing or constrained to depend on borrowed estimates. But even all this may be condoned. Jealousy, vanity, parade of learning, may be pardoned even in a philosopher. Hamilton's deadly sin was none of these, it was sinning against the light. In October, 1877, A. T. Bledsoe, then editor of the *Southern Review* — unfortunately too little known — published an article in that journal in which he proved beyond a reasonable doubt — I have been at the pains to verify the proof — that Hamilton by studied selections and omissions deliberately and maliciously misrepresented the great authors from whom he quoted — d'Alembert, Blaise Pascal, Descartes and others — distorting their express and unmistakable meaning even to the extent of complete inversion. This same verdict regarding Hamilton's vandalism, in so far as it relates to the works of Descartes, was independently reached by Professor Pringsheim and in 1904 announced by him in his *Festrede* before the Munich Academy of Sciences. As for Schopenhauer, I regret to say that a similar charge and finding stand against him also. For not only did he endorse without examination and re-utter Hamilton's tirade in the strongest terms, thus reinforcing it and giving it currency on the continent, but, as Pringsheim has shown, the German philosopher, by careful excision from the writings of Lichtenberg, converts that distinguished physicist's just strictures on the then flourishing but wayward Combinatorial School of mathematics into a severe condemnation of mathematicians in general and of the science itself, which, nevertheless, in the opening but omitted line of the very passage from which Schopenhauer quotes, is characterized by Lichtenberg as *" eine gar herrliche Wissenschaft."* Regarding the question of the intrinsic merit of the estimate of mathematics which these two most famous and influential enemies of the science have made so largely current in the world that it fairly fills the atmosphere and people take it in unconsciously as by a kind of cerebral suction, I shall speak in another connection. What I desire to emphasize here is the fact that neither the vast, splendid, superficial learning of the pompous author of " The Philosophy of the Conditioned " nor the pungence and pith, brilliance and intrepidity of the author of " Die Welt als Wille " can avail to constitute either of them an authority in a subject in which neither was informed and in which both stand convicted falsifiers of the judgments and opinions of other men.

As to Huxley and Holmes, the case is different. Both

of them were generous, genial and honest, and to their opinions on any subject we gladly pay respect qualified only as the former's judgment regarding mathematics was qualified by Sylvester himself:

"Verständige Leute kannst du irren sehn
In Sachen nämlich, die sie nicht verstehn."

In relation to Huxley's statement that mathematical study knows nothing of observation, induction, experiment, and causation, it ought to be borne in mind that there are two kinds of observation: outer and inner, objective and subjective, material and immaterial, sensuous and sense-transcending; observation, that is, of physical things by the bodily senses, and observation, by the inner eye, by the subtle touch of the intellect, of the entities that dwell in the domain of logic and constitute the objects of pure thought. For, phrase it as you will, there is a world that is peopled with ideas, ensembles, propositions, relations, and implications, in endless variety and multiplicity, in structure ranging from the very simple to the endlessly intricate and complicate. That world is not the product but the object, not the creature but the quarry of thought, the entities composing it — propositions, for example, — being no more identical with thinking them than wine is identical with the drinking of it. Mind or no mind, that world exists as an extra-personal affair, — Pragmatism to the contrary notwithstanding. It appears to me to be a radical error of pragmatism to blink the fact that the most fundamental of spiritual things, namely, curiosity, never poses as a maker of truth but is found always and only in the attitude of seeking it. Indeed truth might be defined to be the presupposition or the complement of curiosity — as that without which curiosity would cease to be what it is. The constitution of that extra-personal world, its intimate ontological make-up, is logic in its essential character and substance as an independent and extra-personal form of being, while the study of that constitution is logic pragmatically, in its character, *i.e.*, as an enterprise of mind. Now — and this is the point I wish to stress — just as the astronomer, the physicist, the geologist, or other student of objective science looks abroad in the world of sense, so, not metaphorically speaking but literally, the mind of the mathematician goes forth into the universe of logic in quest of the things that are there; exploring the heights and depths for facts — ideas, classes, relationships, implications, and the rest; observing the minute and elusive with the powerful microscope of his Infinitesimal Analysis; observing the elusive and vast with the limitless telescope of his Calculus of the Infinite; making guesses regarding the order and internal harmony of the data observed and collocated; testing the hypotheses, not merely by the complete induction peculiar to mathematics, but, like his colleagues of the outer world, resorting also to experimental tests and incomplete induction; frequently finding it necessary, in view of unforeseen disclosures, to abandon a once hopeful hypothesis or to transform it by retrenchment or by enlargement: — thus, in his own domain, matching, point for point, the processes, methods and experience familiar to the devotee of natural science.

Is it replied that it was not observation of the objects of pure thought but the other kind, namely, sensuous observation, that Huxley had in mind, then I rejoin that, nevertheless, observation by the inner eye of the things of thought *is* observation, not less genuine, not less difficult, not less rich in its objects and disciplinary value, than is sensuous observation of the things of sense.

But this is not all, nor nearly all. Indeed for direct beholding, for immediate discerning, of the things of mathematics there is none other light but one, namely, psychic illumination, but mediately and indirectly they are often revealed or at all events hinted by their sensuous counterparts, by indications within the radiance of day, and it is a great mistake to suppose that the mathetic spirit elects as its agents those who, having eyes, yet see not the things that disclose themselves in solar light. To facilitate eyeless observation of his sense-transcending world, the mathematician invokes the aid of physical diagrams and physical symbols in endless variety and combination; the logos is thus drawn into a kind of diagrammatic and symbolical incarnation, gets itself externalized, made flesh, so to speak; and it is by attentive physical observation of this embodiment, by scrutinizing the physical frame and make-up of his diagrams, equations and formulae, by experimental substitutions in, and transformations of, them, by noting what emerges as essential and what as accidental, the things that vanish and those that do not, the things that vary and the things that abide unchanged, as the transformations proceed and trains of algebraic evolution unfold themselves to view, — it is thus, by the laboratory method, by trial and by watching, that often the mathematician gains his best insight into the constitution of the invisible world thus depicted by visible symbols. Indeed the importance to the mathematician of such sensuous observation cannot be overrated. It is not merely that the craving to see has led to the construction of the manifold models, ingenious and noble, of Schilling and others, illustrating important parts of Higher Geometry, Analysis Situs, Function Theory and other doctrines, but the annals of the science are illustrious with achievements made possible by facts first noted by the physical eye. To take a simple example from ancient days, it was by observation of the fact that the squares of certain numbers are each the sum of two other squares, the detection and collection of these numbers by the method of trial, observation of the fact that apparently all and only the numbers of such triplets are measures of the sides of right triangles, — it was thus, by observation and experiment, by the method of incomplete induction, common to the experimental sciences, that the Pythagorean theorem, now familiar throughout the world, was discovered. It was by Leibniz's observation of the definitely lawful manner in which the coefficients of a system of equations enter their solution that the suggestion came of a notion on the basis of which there has grown up in our time an imposing theory, an algebra built up on algebra — the colossal doctrine of Determinants. It was the observation, the detection by the eye of Lagrange and Boole and Eisenstein, of the fact that linear transformation of certain algebraic expressions leaves certain functions of their coefficients absolutely undisturbed in form, unaltered in frame of constitution, that gave rise to the concept, and therewith to the morphological doctrine, of Invariants, a theory filling the heavens like a light-bearing ether, penetrating all the branches of geometry and analysis, revealing everywhere abiding configurations in the midst of change, everywhere disclosing the eternal reign of the law of Form. It was in order to render evident to sensuous observation and to keep constantly before the physical eye the pervasive symmetry of mathematical thought that Hesse in the employment of homogeneous coördinates set the example, since then generally followed, of replacing a variety of different letters by repetitions of a single one distinguished by indices or subscripts, — a practice yet further justified on grounds both of physical and of intellectual economy. It was by sensuous observation that Clerk Maxwell, in the beginning of his wondrous career, detected a lack of symmetry in the then recognized equations of electro-dynamics and by that observed fact together with a discriminating sense of the scientific significance of esthetic intimations, that he was led to remove the seeming blemish by the addition of a term, antedating experimental justification of his daring deed by twenty years: an example of prescience not surpassed by that of Adams and Leverrier who, while engaged in the study of planetary disturbance, each of them about the same time and independently of the other, felt the then unknown Neptune " trembling on the delicate thread of their analysis " [8] and correctly informed the astronomer where to point his telescope in order to behold the planet. One might go on to cite the theorem of Sturm in Equation Theory, the " Diophantine theorems of Fermat " in the Theory of Numbers, the Jacobian " doctrine of double periodicity " in Function Theory, Legendre's law of reciprocity, Sylvester's reduction of Euler's problem of the Virgins to the form of a question in Simple Partitions, and so on and on, thus continuing indefinitely the story of the great rôle of observation, experiment and incomplete induction, in mathematical discovery. Indeed it is no wonder that even Gauss, " facile princeps matematicorum," even though he dwelt aloft in the privacy of a genius above the needs and ways of other minds, yet pronounced mathematics " a science of the eye."

Indeed the time is at hand when at least the academic mind should discharge its traditional fallacies regarding

[8] Quotation from an essay by Professor W. B. Smith.

the nature of mathematics and thus in a measure promote the emancipation of criticism from inherited delusions respecting the kind of activity in which the life of the science consists. Mathematics is no more the art of reckoning and computation than architecture is the art of making bricks or hewing wood, no more than painting is the art of mixing colors on a palette, no more than the science of geology is the art of breaking rocks, or the science of anatomy the art of butchering.

Did not Babbage or somebody invent an adding machine? And does it not follow, say Holmes and Schopenhauer, that mathematical thought is a merely mechanical process? Strange how such trash is occasionally found in the critical offering of thoughtful men and thus acquires circulation as golden coin of wisdom. It would not be sillier to argue that, because Stanley Jevons constructed a machine for producing certain forms of logical inference, therefore all thought, even that of a philosopher like Schopenhauer or that of a poet like Holmes, is merely a thing of pulleys and levers and screws, or that the pianola serves to prove that a symphony by Beethoven or a drama by Wagner is reducible to a trick of mechanics.

But far more pernicious, because more deeply imbedded and persistent, is the fallacy that the mathematician's mind is but a syllogistic mill and that his life resolves itself into a weary repetition of *A* is *B, B* is *C,* therefore *A* is *C;* and *Q.E.D.* That fallacy is the *Carthago delenda* of regnant methodology. Reasoning, indeed, in the sense of compounding propositions into formal arguments, is of great importance at every stage and turn, as in the deduction of consequences, in the testing of hypotheses, in the detection of error, in purging out the dross from crude material, in chastening the

deliverances of intuition, and especially in the final stages
of a growing doctrine, in welding together and con-
catenating the various parts into a compact and coherent
whole. But, indispensable in all such ways as syllogistic
undoubtedly is, it is of minor importance and minor
difficulty compared with the supreme matters of Invention and Construction. *Begriffbildung,* the resolution of
the nebula of consciousness into star-forms of definite
ideas; discriminating sensibility to the logical signifi-
cances, affinities and bearings of these; susceptibility to
the delicate intimations of the subtle or the remote;
sensitiveness to dim and fading tremors sent below by
breezes striking the higher sails; the ability to grasp
together and to hold in steady view at once a multitude
of ideas, to transcend the individuals and, compounding
their forces, to seize the resultant meaning of them all;
the ability to summon not only concepts but doctrines,
marshalling them and bringing them to bear upon a
single point, like great armies converging to a critical
center on a battle field. These and such as these are the
powers that mathematical activity in its higher rôles
demands. The power of ratiocination, as already said,
is of exceeding great importance but it is neither the base
nor the crown of the faculties essential to " Mathema-
ticised Man." When the greatest of American logicians,
speaking of the powers that constitute the born geometrician, had named Conception, Imagination, and
Generalization, he paused. Thereupon from one in the
audience there came the challenge, " What of Reason? "
The instant response, not less just than brilliant, was
" Ratiocination — that is but the smooth pavement on
which the chariot rolls." When the late Sophus Lie,
great comparative anatomist of geometric theories,
creator of the doctrines of Contact Transformations, and

Infinite Continuous Groups, and revolutionizer of the Theory of Differential Equations, was asked to name the characteristic endowment of the mathematician, his answer was the following quaternion: *Phantasie, Energie, Selbstvertrauen, Selbstkritik.* Not a word, you observe, about ratiocination. *Phantasie,* not merely the fine frenzied fancy that gives to airy nothings a local habitation and a name, but the creative imagination that conceives ordered realms and lawful worlds in which our own universe is as but a point of light in a shining sky; *Energie,* not merely endurance and doggedness, not persistence merely, but mental *vis viva,* the kinetic, plunging, penetrating power of intellect; *Selbstvertrauen* and *Selbstkritik,* self-confidence aware of its ground, deepened by achievement and reinforced until in men like Richard Dedekind, Bernhard Bolzano and especially Georg Cantor it attains to a spiritual boldness that even dares leap from the island shore of the Finite over into the all-surrounding boundless ocean of Infinitude itself, and thence brings back the gladdening news that the shoreless vast of Transfinite Being differs in its logical structure from that of our island home only in owning the reign of more *generic law.*

Indeed it is not surprising, in view of the polydynamic constitution of the genuinely mathematical mind, that many of the major heroes of the science, men like Desargues and Pascal, Descartes and Leibniz, Newton, Gauss, and Bolzano, Helmholtz and Clifford, Riemann and Salmon and Plücker and Poincaré, have attained to high distinction in other fields not only of science but of philosophy and letters too. And when we reflect that the very greatest mathematical achievements have been due, not alone to the peering, microscopic, histologic vision of men like Weierstrass, illuminating the hidden recesses, the minute and intimate structure of logical reality, but to the larger vision also of men like Klein who survey the kingdoms of geometry and analysis for the endless variety of things that flourish there, as the eye of Darwin ranged over the flora and fauna of the world, or as a commercial monarch contemplates its industry, or as a statesman beholds an empire; when we reflect not only that the Calculus of Probability is a creation of mathematics but that the master mathematician is constantly required to exercise judgment — judgment, that is, in matters not admitting of certainty — balancing probabilities not yet reduced nor even reducible perhaps to calculation; when we reflect that he is called upon to exercise a function analogous to that of the comparative anatomist like Cuvier, comparing theories and doctrines of every degree of similarity and dissimilarity of structure; when, finally, we reflect that he seldom deals with a single idea at a time, but is for the most part engaged in wielding organized hosts of them, as a general wields at once the divisions of an army or as a great civil administrator directs from his central office diverse and scattered but related groups of interests and operations; then, I say, the current opinion that devotion to mathematics unfits the devotee for practical affairs should be known for false on *a priori* grounds. And one should be thus prepared to find that as a fact Gaspard Monge, creator of descriptive geometry, author of the classic " Applications de l'analyse à la géométrie "; Lazare Carnot, author of the celebrated works, " Géométrie de position," and " Réflexions sur la Métaphysique du Calcul infinitesimal "; Fourier, immortal creator of the " Théorie analytique de la chaleur "; Arago, rightful inheritor of Monge's chair of geometry; and Poncelet, creator of pure projective

geometry; one should not be surprised, I say, to find that
these and other mathematicians in a land sagacious
enough to invoke their aid, rendered, alike in peace and in
war, eminent public service.

To speak at length, if that were necessary, of Huxley's
deliverance that the study of mathematics " knows nothing of causation," the " law of my song and the hasten-
ing hour forbid." Suffice it to say in passing that when
the mathematician seeks the consequences of given sup-
positions, saying ' when these precede, those will follow,'
and when, having plied a circle, a sphere or other form
chosen from among infinitudes of configurations, with
some transformation among infinite hosts at his disposal,
he speaks of its ' effect,' then, I submit, he is employing
the language of causation with as nice propriety as it
admits of in a world where, as everyone knows, except
such as still enjoy the blessings of a juvenile philosophy,
the best we can say is that the ceaseless shuttles fly back
and forth, and streams of events without original source
flow on without ultimate termination. Indeed it is a
certain and signal lesson of science in all its forms everywhere that the language of cause and effect, except in
the sense of facts being lawfully implied in other facts,
has no indispensable use.

I have not spoken of " Applied Mathematics," and that
for the best of reasons: there is, strictly speaking, no
such thing. The term indeed exists, and, in a conserva-
tive practical world that cares but little for " The nice
sharp quillets of the law," it will doubtless persist as a
convenient designation for something that never existed
and never can. It is of the very essence of the practician
type of mind not to know aught as it is in itself nor
aught as self-justified but to mistake the secondary and
accidental for the primary and essential, to blink and

elude the presence of *immediate* worth, and being thus
blind to instant and immanent ends, to revel in means
and uses and applications, requiring all things to excuse
their being by extraneous and emanant effects, — vindi-
cating the stately elm by its promise of lumber, or the
lily by its message of purity, or the flood of Niagara by
its available energy, or even knowledge itself by the
worldly advantage and the power which it gives. I am
told that even the deep and exquisite terminology of art
has been to some extent invaded by such barbarous and
shallow phrases as 'applied music,' 'applied architecture,' 'applied sculpture,' 'applied painting,' as if
Beauty, virgin mother of art, could, without dissolution
of her essential character, consciously become the willing
drudge and paramour of Use. And I suppose we are
fated yet to hear of applied glory, applied holiness,
applied poetry — *i.e.*, poetry that is consciously peda-
gogic or that aims at a moral and thereby sinks or rises
to the level of a sermon — of applied joy, applied ontology, yea, of applied inapplicability itself.

It is in implications and not in applications that mathematics has its lair. Applied mathematics is mathematics
simply or is not mathematics at all. To think aright is
no characteristic striving of a class of men; it is a common aspiration; and Mechanics, Mathematical Physics,
Mathematical Astronomy, and the other chief *Anwen-
dungsgebiete* of mathematics, as Geodesy, Geophysics,
and Engineering in its various branches, are all of them
but so many witnesses of the truth of Riemann's saying
that " Natural science is the attempt to comprehend
nature by means of exact concepts." A gas molecule
regarded as a minute sphere or other geometric form,
however complicate; stars and planets conceived as
ellipsoids or as points, and their orbits as loci; time and

space, mass and motion and impenetrability; velocity, acceleration and energy; the concepts of norm and average; — what are these but mathematical notions? And the wondrous garment woven of them in the loom of logic — what is that but mathematics? Indeed every branch of so-called applied mathematics is a mixed doctrine, being thoroughly analyzable into two disparate parts: one of these consists of determinate concepts formally combined in accordance with the canons of logic, *i.e.*, it is mathematics and not natural science viewed as matter of observation and experiment; the other *is* such matter and is natural science in that conception of it and not mathematics. No fibre of either component is a filament of the other. It is a fundamental error to regard the term Mathematicisation of thought as the importation of a tool into a foreign workshop. It does not signify the transition of mathematics conceived as a thing accomplished over into some outlying domain like physics, for example. Its significance is different radically, far deeper and far wider. It means the growth of mathematics itself, its extension and development from within; it signifies the continuous revelation, the endlessly progressive coming into view, of the static universe of logic; or, to put it dynamically, it means the evolution of intellect, the upward striving and aspiration of thought everywhere, to the level of cogency, precision and exactitude. This self-propagation of the rational logos, the springing up of mathetic rigor even in void and formless places, in the very retreats of chaos, is to my mind the most impressive and significant phenomenon in the history of science, and never so strikingly manifest as in the last half hundred years. Seventy-two years ago, even Comte, the stout advocate of mathematics as constituting " the veritable point of departure for all rational scientific education, general or special," expressed the opinion that we should never " be in position by any means whatever to study the chemical composition of the stars." In less than twenty-five years thereafter that negative prophecy was falsified by the chemical genius of Bunsen fortified by the mathematics of Kirchoff. Not only has mathematics grown, in the domain of Physics, into the vast proportions of Rational Dynamics, but the derivative and integral of the Calculus, and Differential Equations, are more and more finding subsistence in Chemistry also, and by the work of Nernst and others even the foundations of the latter science are being laid in mathematico-physical considerations. Merely to sketch most briefly the mathematical literature that has grown up in the field of Political Economy requires twenty-five pages of the above mentioned *Encyklopädie* of mathematics. Similar sketches for Statistics and Life Insurance require no less than thirty and sixty-five pages respectively. Even in the baffling and elusive matter of Psychology, the work of Herbart, Fechner, Weber, Wundt and others confirms the hope that the soil of that great field will some day support a vigorous growth of mathematics. It seems indeed as if the entire surface of the world of human consciousness were predestined to be covered over, in varying degrees of luxuriance, by the flora of mathetic science.

But while mathematics may spring up and flourish in any and all experimental and observational fields, it is by no means to be expected that ' experiment and observation ' will ever thus be superseded. Such domains are rather destined to be occupied at the same time by two tenants, mathematical science and science that is not mathematical. But while the former will serve as an ideal standard for the latter, mathematics has neither the power nor the disposition to disseize experiment and observation of any holdings that are theirs by the rights of conquest and use. Between mathematics on the one hand and non-mathematical science on the other, there can never occur collision or quarrel, for the reason that the two interests are ultimately discriminated by the kind of curiosity whence they spring. The mathematician is curious about definite naked relationships, about logically possible modes of order, about varieties of implication, about completely determined or determinable functional relationships, considered solely in and of themselves, considered, that is, without the slightest concern about any question whether or no they have any external or sensuous validity or other sort of validity than that of being logically thinkable. It is the aggregate of things thinkable logically that constitutes the mathematician's universe, and it is inconceivably richer in mathetic content than can be any outer world of sense such as the physical universe according to which we chance to have our physical being.

This mere speck of a physical universe in which the chemist, the physicist, the astronomer, the biologist, the sociologist, and the rest of nature students, find their great fields and their deep and teeming interests, may be a realm of invariant uniformities, or laws; it may be a mechanically organic aggregate, connected into an ordered whole by a tissue of completely definable functional relationships; and it may not. It may be that the universe eternally has been and is a genuine cosmos; it may be that the external sea of things immersing us, although it is ever changing infinitely, changes only lawfully, in accordance with a system of immutable rules of order that constitute an invariant at once underived and indestructible and securing everlasting harmony through and through; and it may not be such. The student of nature assumes, he rightly assumes, that it is; and, moved and sustained by characteristic appropriate curiosity, he endeavors to find in the outer world what are the elements and what the relationships assumed by him to be valid there. The mathematician as such does not make that assumption and does not seek for elements and relationships in the outer world.

Is the assumption correct? Undoubtedly it is admissible, and as a working hypothesis it is undoubtedly exceedingly useful or even indispensable to the student of external nature; but is it true? The mathematician as man does not know although he cares. Man as mathematician neither knows nor cares. The mathematician does know, however, that, if the assumption be correct, every relationship that is valid in nature is, *in abstractu*, an element in his domain, a subject for his study. He knows, too, at least he strongly suspects, that, if the assumption be not correct, his domain remains the same absolutely, and the title of mathematics to human regard " would remain unimpeached and unimpaired " were the universe without a plan or, having a plan, if it " were unrolled like a map at our feet, and the mind of man qualified to take in the whole scheme of creation at a glance."

The two realms, of mathematics, of natural science, like the two curiosities and the two attitudes, the mathematician's and the nature student's, are fundamentally distinct and disparate. To think logically the logically thinkable — that is the mathematician's aim. To assume that nature is thus thinkable, an embodied rational logos, and to discover the thought supposed incarnate there — these are at once the principle and the hope of the student of nature.

Suppose the latter student is right and that the outer universe really is an embodied logos of reason, does it follow that all the logically thinkable is incorporated in it? It seems not. Indeed there appears to be many a rational logos. A cosmos, a harmoniously ordered universe, one that through and through is self-compatible, can hardly be the whole of reason materialized and objectified. At all events the mathematician has delight in the conceptual construction and in the contemplation of divers systems that are inconsistent with one another though each is thoroughly self-coherent. He constructs in thought a summitless hierarchy of hyperspaces, an endless series of ordered worlds, worlds that are possible and logically actual. And he is content not to know if any of them be otherwise actual or actualized. There is, for example, a Euclidean geometry and there are infinitely many kinds of non-Euclidean. These doctrines, regarded as *true* descriptions of some one actual space, are incompatible. In our universe, to be specific, if it be as Plato thought and natural science takes for granted, a geometrized or geometrizable affair, then one of these geometries may be, none of them may be, not all of them can be, objectively valid. But in the infinitely vaster world of pure thought, in the world of mathesis, all of them are valid; there they co-exist, there they interlace and blend among themselves and others as differing strains of a hypercosmic harmony.

It is from some such elevation, not the misty lowland of the sensuously and materially Actual, but from a mount of speculation lawfully rising into the azure of the logically Possible, that one may glimpse the dawn heralded by the avowal of Leibniz: " *Ma métaphysique est toute mathématique.*" Time fails me to deal fittingly with the great theme herewith suggested, but I cannot

quite forbear to express briefly my conviction that, apart
from its service to kindred interests of thought as a
standard of clarity, rigor and certitude, mathematics is
and will be found to be an inexhaustible quarry of material — of ontologic types, of ideas and problems, of distinctions, discriminants and hints, evidences, analogies
and intimations — all for the exploitation and use of Philosophy, Psychology, and Theology. The allusion is not
to such celebrated alliances of philosophy and mathesis
as flourished in the school of Pythagoras and in the gigantic personalities of Plato, Descartes, Spinoza, and
Leibniz, nor to the more technical mathematico-philo-
sophical researches and speculations of our own time by
such as C. S. Peirce, Russell, Whitehead, Peano, G.
Cantor, Couturat and Poincaré, glorious as were those
alliances and important as these researches are. The
reference is rather to the unappreciated fact that the
measureless accumulated wealth of the realm of exact
thought is at once a marvelous mine of subject matter
and a rich and ready arsenal for those great human concerns of reflective and militant thought that is none the
less important because it is not exact.

For the vindication of that claim, a hint or two must
here suffice. The modern mathematical concepts of
number, time, space, order, infinitude, finitude, group,
manifold, functionality, and innumerable hosts of others,
the varied processes of mathematics, and the principles
and modes of its growth and evolution, all of these or
nearly all still challenge and still await those kinds of
analysis that are proper to the philosopher and the psychologist. The psychology of Euclidean, non-Euclidean,
and hyperspaces, the question of the intuitability of the
latter, the secret of their having become not only indispensable in various branches of mathematics but instru-

mentally useful in other fields also, as in the kinetic theory of gases; the question, for example, why it is that while *thought* maintains a straightforward course through four-dimensional space, *imagination* travels through it on a zigzag path, of two logically identical configurations, being partially or completely blind to the one, yet perfectly beholding the other; the evaluation and adjustment of the contradictory claims of Poincaré and his school on the one hand and of Mach and his disciples on the other, the former contending that Modern Analysis is a " free creation of the human spirit " guided indeed but not constrained by experience of the external world, being merely kept by this from aimless wandering in wayward paths; while the latter maintain that mathematical concepts, however tenuous or remote or recondite, have been literally evolved continuously in accordance with the needs of the animal organism and with environmental conditions out of the veriest elements (feelings) of physical life, and accordingly that the purest offspring of mathematical thought may trace a legitimate lineage back and down to the lowliest rudiments of physical and physiological experience: — these problems and such as these are, I take it, problems for the student of mind as mind and for the student of psycho-physics.

Regarding the relations of mathesis to the former "queen of all the sciences," I have on this occasion but little to say. I do not believe that the declined estate of Theology is destined to be permanent. The present is but an interregnum in her reign and her fallen days will have an end. She has been deposed mainly because she has not seen fit to avail herself promptly and fully of the dispensations of advancing knowledge. The aims, however, of the ancient mistress are as high as ever, and when she shall have made good her present lack of modern education and learned to extend a generous and eager hospitality to modern light, she will reascend, and will occupy with dignity as of yore an exalted place in the ascending scale of human interests and the esteem of enlightened men. And mathematics, by the character of her inmost being, is especially qualified, I believe, to assist in the restoration. It was but little more than a generation ago that the mathematician, philosopher and theologian, Bernhard Bolzano, dispelled the clouds that throughout all the foregone centuries had enveloped the notion of Infinitude in darkness, completely sheared the great term of its vagueness without shearing it of its strength, and thus rendered it forever available for the purposes of logical discourse. Whereas, too, in former times the Infinite betrayed its presence not indeed to the faculties of Logic but only to the spiritual Imagination and Sensibility, mathematics has shown, even during the life of the elder men here present, — and the achievement marks an epoch in the history of man, — that the structure of Transfinite Being is open to exploration by the organon of Thought. Again, it is in the mathematical doctrine of Invariance, the realm wherein are sought and found configurations and types of being that, amid the swirl and stress of countless hosts of transformations, remain immutable, and the spirit dwells in contemplation of the serene and eternal reign of the subtile law of Form, it is there that Theology may find, if she will, the clearest conceptions, the noblest symbols, the most inspiring intimations, the most illuminating illustrations, and the surest guarantees of the object of her teaching and her quest, an Eternal Being, unchanging in the midst of the universal flux.

It is not, however, by any considerations or estimates of utility in any form however high it be or essential to

the worldly weal of man; it is not by evaluating mastery
of the processes of measurement and computation,
though these are continuously vital everywhere to the
conduct of practical life; nor is it by strengthening the
arms of natural science and speeding her conquests in
a thousand ways and a hundred fields; nor yet by extending the empire of the human intellect over the realms
of number and space and establishing the dominion of
thought throughout the universe of logic; it is not even
by affording argument and fact and light to theology and
so contributing to the advancement of her supreme concerns; — it is not by any of these considerations nor by
all of them that Mathematics, were she called upon to
do so, would rightly seek to vindicate her highest claims
to human regard. It requires indeed but little penetration to see that no science, no art, no doctrine, no human
activity whatever, however humble or high, can ultimately succeed in justifying itself in terms of measurable
fruits and emanant effects, for these remain always to
be themselves appraised, and the process of such attempted vindication is plainly fated to issue only in re-
gression without an end. Such Baconian apologetic,
when offered as final, quite mistakes the finest mood of
the scientific spirit and is beneath the level of academic
faith. Science does not seek emancipation in order to
become a drudge, she consents to serve indeed but her
service aims at freedom as an end.

Man has been so long a slave of circumstance and
need, he has been so long constrained to seek license for
his summit faculties, in lower courts without appeal, that
a sudden transitory moment of release sets him trembling
with distrust and fear, an occasional imperfect vision of
the instant dignity of his spiritual enterprises is at once
obscured by doubt, and he straightway descends into the

market places of the world to excuse or to justify his illumination, pleading some mere utility against the ignoring or the condemnation of an insight or an inspiration whose worth is nevertheless immediate and no more needs and no more admits of utilitarian justification than the breaking of morning light on mountain peaks or the bounding of lambs in a meadow.

The solemn cant of Science in our day and her sombre visage are but the lingering tone and shade of the prison-house, and they will pass away. Science is destined to appear as the child and the parent of freedom blessing the earth without design. Not in the ground of need, not in bent and painful toil, but in the deep-centred play-instinct of the world, in the joyous mood of the eternal Being, which is always young, Science has her origin and root; and her spirit, which is the spirit of genius in moments of elevation, is but a sublimated form of play, the austere and lofty analogue of the kitten playing with the entangled skein or of the eaglet sporting with the mountain winds.

MAN AND MEN [1]

I CHOSE this subject — " Man and Men " — because I desired to discuss with you the most important subject that I or you or anyone else could think of. In our world there are many realities but they differ much in dignity or rank. Of all the realities with which you and I have to deal, with which it is our privilege and obligation to deal, the supreme one is not matter nor material energy nor space nor time, though the importance of these is very great. The supreme reality is Man. The supreme concrete realities of our world are human individuals — men, women, children. The supreme abstract reality of the world is man — the human race — Humanity. What do I mean by that term? I mean, I suppose, what you mean. By Humanity I mean all mankind — not merely the living — but the living, the dead, and the unborn. By Humanity I mean, if I may answer abstractly, those propensities and powers in virtue of which humans are *human*. Have you ever considered what those propensities and powers are? I shall not tarry here, for it would detain us too long, to name them and analyze them. Being human, you have them in some measure. As *intelligent* humans, interested to understand your own nature, you are bound to ascertain what they are, if you can. And you can, if you will. There is a book that will greatly help you in the quest.

[1] Published in the *Builder,* Jan., 1925. It is an extract from an address delivered before the Bureau of Personnel Administration, New York, in January, 1924.

I mean Count Korzybski's *Manhood of Humanity*. It is a work that you and every other man and woman ought to read open-mindedly, re-read, and ponder. With its central idea I have dealt briefly in a chapter of my *Mathematical Philosophy*, but the reading of that chapter, though it may help, is far from sufficient.

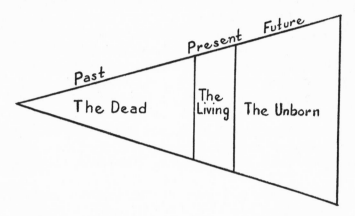

In the remaining minutes of the hour I wish to say a few things, by way of opening up an immense subject, about the relations of individual human beings to Humanity. I can hardly do more than drive into the wall, rather rudely, a few wooden pegs to which you may desire to attach some reading and reflection in the future.

The diagram given herewith will help us. The mid-part represents the Present, occupied by the existing people of the world. On the left is our human Past, tenanted by the dead, a long backward stretch, perhaps a half million years. On the right is the wide strange Future, mysterious realm of the unborn. Humanity embraces the three — the Past, the Present, the Future —

the dead, the living, the unborn. Today you and I are among the living. Yesterday we were unborn. To-morrow our bodies will have perished. In each estate we are members of Humanity — representatives, as W. K. Clifford might have said, of " Father Man."

We are here tonight because it is the fortune of the living to have to deal with what we are wont to call " the problems of the world." Of these the so-called industrial problems, in which you are especially interested, are only some of the elements or aspects. All the problems are primarily and essentially human problems, problems about humans for humans to solve, and no such problem exists alone or admits of a lone solution. Each of them is in a network involving all the others. In this matter the mystic's contention is true — each is all and all is each.

As Maggots in a Cheese

I have no single formula for the solution of all your industrial problems, much less have I one for solving all the great problems of our troubled modern world. But it is my conviction that the chief source of trouble is this: *We have been and are living in the midst of a great civilization like maggots in a cheese.*

That is not a conclusion arrived at in haste. It is very deliberate. Perhaps you will consider it carefully.

We are living immersed in a civilization which, despite all its short-comings, is so vast and rich and manifold that we cannot measure its proportions nor assess its worth. The industrial and other social troubles of the world will be found, if we view them fundamentally, to have their *roots* in the fact that we have been living, and are still living, in and upon that civilization, without

serious thought of our relations to it, without a sense of
our indebtedness and obligations, like maggots in a
cheese.

And so the best formula I can think of for dealing
with industrial and the other human problems of the
world is this: *Stop living in the midst of our great modern
civilization as maggots in a cheese.*

How is that to be done? The first means thereto is
to study civilization — its origin, its genesis, its essential
nature, our human relations to it and our obligations.

Where did our civilization come from?

Let me intimate the answer by means of an example.
Some months ago I was teaching a class in the calculus.
The boys were dealing with problems in *maxima* and
minima. Some of you know, and others do not know,
what that means, but all of you will understand what I
am about to say. We had one hour. The boys finished
in forty-five minutes. " Boys," I said, " please be seated.
I want to say something more important than the calculus.

"You and I are probably quite ordinary people. It may be that
some one among you is extraordinary but, if so, the happy fact is not
yet manifest. So let us assume that our native gifts are ordinary.
Yet you have just now readily solved problems of a kind that the
greatest genius that ever lived on our planet could not solve without
the instrument you have employed — an instrument you have been
getting acquainted with during the last few weeks, namely, the calculus.
Where did it come from? "

"It was invented by Newton," said one of the boys. Another one
said: " It was invented by Leibniz."

I said: " Boys, both of those answers are commonly given and in
a sense they are correct. But in a deeper sense both of them are
false because Newton and Leibniz did but improve the mathematics
of their immediate predecessors, and these did but improve the mathematics of *their* predecessors, and so on back till you come down here
somewhere in the sharp angle of our diagram where our dear remote
ancestors are engaged in the struggle of learning to count: the calculus
was invented back there ages ago — I mean it started there. And so
the calculus was not created by Newton or Leibniz. It was produced.

little by little, by many generations now in the state of those we are
accustomed to call the dead."

And I said: " Boys, they are not dead — that must be evident to
you. Their bodies are dead but the men are living and are here in
this room. Newton and Leibniz and the rest are here — they are at
work, working *with* us and *through* us as agents of Humanity, by
means of ideas which they invented, which we inherited and which it
is our privilege as humans, and our obligations, to use, to improve,
and to transmit. I say ' transmit,' for the unborn are coming — if you
will go to your cloister and there meditate in the silence, you can see
them approach, generation after generation of them, fellow children
with us of ' Father Man ' — they look to you and me and appeal to
us as the present occupants and guardians of their future home, for
the kind of world they will find depends upon our loyalty as representatives of Humanity."

THE PRINCIPLE IS UNIVERSALLY TRUE

I have used the calculus, my friends, merely as an illustration. The calculus is but one element of our civilization. What I have said of the calculus is true of all
the other elements — speech, the arts, the sciences, the
inventions, the great literatures of East and West, the
wisdoms of philosophy and law, the ways of social or-
ganization and order, and all the other kinds and forms
of material and spiritual wealth. We, the people of this
generation, were born in the midst of an immense civilization. We may have improved it a little in some respects
but we did *not create* it. It is of the utmost importance
for us to grasp that fact and hold it fast and realize it
keenly; for else we shall be as maggots in a cheese. Our
civilization — the material and spiritual wealth of the
world — was not produced by us. We have it as a gift.
It is the fruit of the time and thought and toil of many
generations of those whose bodies have indeed perished
but whose spirits survive and are now active in the ideas
and ideals and sentiments and aspirations embodied and

transmitted to us in the form of instrumentalities and institutions, knowledges and arts.

We must understand and not forget that there was a time when there were no human beings on this globe. There was a time when humans began to be. We must try to realize, for it is true, that our remotest human ancestors did not know what they were nor where they were. They had no clothes nor houses — they were probably covered with hair and dwelt in caves. They had no language, no human history, not even human tradition, no knowledge of number, no guiding maxims, no tools nor craftmanship. But they had a marvelous thing — a gift that enabled them and impelled them to *start* what *we* call civilization; and they were, moreover, the first of a race that had another equally marvelous and equally precious gift — a gift enabling them and impelling them to *advance* civilization. These are the gifts that make humans human.

Civilization is the Creature of Man

And so we see that civilization is the creature, not of men, but of Man. It is the product of Humanity. It is to the time and thought and toil of those remote rude ancestors, groping in the dark, and of the many generations of their descendants that you and I and our living fellows are indebted for the immeasurable riches — the material and spiritual wealth — of our present world.

To receive that Human Inheritance as we habitually do receive it, taking it all for granted as we take the gifts of Physical Nature — land and light and sea and sky; not to realize in thought and in feeling that, though we are individuals, we are living organs of Humanity; not to realize in our heads and hearts and ways of living, and not to teach in home and school, our relations and obligations to the Dead and the Unborn: *that* is what I mean by "living in and upon our civilization like maggots in a cheese."

But in proportion as we learn to understand and to feel those relations and obligations, we shall emancipate ourselves from the lower ideals dominant in the world and come undèr the sway of the higher ones. For, as Benjamin Kidd has justly insisted, there is a hierarchy of ideals and a hierarchy of emotions begotten of them. From the power of the emotion of the ideal of self-efficiency — causing us to live and kill and die for *self;* from the power of the emotion of the tribal ideal — causing us to live and kill and die for *tribe;* from the power of the emotion of the state ideal — causing us to live and kill and die for *state:* from the domination of these we shall emancipate ourselves and more and more come under the sway of the highest of all possible ideals, the ideal of Man — causing us to live and, without killing, to die for Humanity.

EDUCATIONAL IDEALS THAT ARE MOST WORTHY OF LOYALTY[1]

THOSE who are engaged in acquiring education ought to be interested in problems of education no less than are those who are engaged in the work of instruction or in that of educational administration. I desire to speak to you briefly regarding an educational problem that is distinguished, I believe, by the following marks; among educational problems it is of supreme importance; it is a problem that presents itself to each of us individually and not merely to those who, in one way or another, represent us; and the solution of the problem is, if I am not mistaken, the same for all. I refer to the problem of ideals. I refer to the problem of selecting those ideals that are most worthy of our devotion and our loyalty, for ideals, which are the sources of light for our pathways, are not of equal worth. They differ in respect of dignity and importance, differing, like the stars, in glory.

A few weeks ago as I was walking one night down Bancroft Way I had the fortune to be joined by a young man who came out from his residence as I was passing the gate, and who, though neither of us knew the other, greeted me in good western fashion. This young man was an ingenuous youth and I soon learned that he was a student in the University of California. In the course of our conversation and in response to some questions,

[1] Stenographic report of an address at the university meeting of the University of California, October 27, 1916.

he informed me that before coming hither as a student he had definitely decided what was to be his pursuit in life and that he was confining his attention to those courses of study that bore most immediately upon it, neglecting all other courses and all other subjects. This excellent young man was, in my judgment, making a grave mistake. The mistake did not consist in his having chosen a pursuit, for that is a choice which all of us must make sooner or later. Neither did the mistake consist in his resolving to equip himself thoroughly for his vocation, for such equipment is indispensable in a world where, in all occupations, competition is keen and is destined to become keener and keener with the passing of the years. His mistake consisted in his forgetting that he was a man and in remembering only that he was to be a follower of a pursuit. And in thus forgetting and remembering, he forgot what was major and remembered only what was minor; he remembered what was subordinate and forgot what was supreme. For man is infinitely superior to any specific form of human activity: to be a man is to be something immeasurably greater than to be the most successful follower of any pursuit whether it be medicine, or law, or theology, or teaching, or agriculture, or any other specific variety among the callings of men. As animate beings inhabiting a world where humans, like the animals, are obliged to win their life from day to day, we, all of us, are or are destined to be, in one way or another, hewers of wood and drawers of water. On this account we all of us require what is called vocational or professional training. The ideal of such education is efficiency. But as men and women, as representatives of the race of man, we are called to something higher; we are called to that kind of education which has been known for more than two thousand years and which I hope will continue to be known under the beautiful designation of liberal education. The ideal of liberal education is not mere efficiency. Its ideal is intelligence, emancipation, magnanimity. It is intelligence because intelligence is the sole means to emancipation. It is emancipation because emancipation is the sole means to magnanimity. It is magnanimity because magnanimity is the highest estate of man. I desire to warn you, as your friend, against the enemies of liberal education. These are very numerous, being easy to produce, springing up like weeds along the dusty highway, almost under the very hoof of travel. I desire to warn you against all manner of practicianism. I desire to warn you against the insidious and baleful influence of those omnipresent, well-meaning, wingless-minded educators who unconsciously conceive young men and women as more or less sublimated beasts and who regard colleges and universities as agencies for teaching the animals the arts of getting shelter and raiment and food.

I have said that the ideal of liberal education is not mere efficiency but is intelligence, emancipation, magnanimity. Accordingly it is the function of liberal education to orient and discipline our human faculties, not merely in their relation to the nature of some pursuit, but in their relation to all the great permanent massive facts of life and the world. I have not time to argue that one who aspires to the kind of education that is appropriate to us in our characters as men and women, may not neglect the discipline afforded by history and the literature of antiquity, because one of the great abiding facts of life and the world is the fact that each of us has behind him an immense human past, a past of which we are children, a past which holds for our guidance and edification the record or the ruins of all the experiments that man has made in many thousands of years in the art of living in this world. I have not time to argue that those who hope to attain, in the course of life, to the spiritual status of a liberally educated mind, may not neglect the discipline of natural science, for the reason that one of the great abiding massive facts of life and the world is the fact that we humans are completely immersed in an infinite universe of matter and force, of which we are literally parts and in regard to which it is the function of science to give us intelligence. I have not time to argue that one devoted to acquisition of liberal education may not neglect the discipline of political or social science and jurisprudence for one of the invariant cardinal facts of life and the world is the fact that man, said Aristotle, is made for cooperation, being by nature a social creature born to membership in a thousand teams in which he must work or perish. I have not time to argue that an aspirant to liberal education may not neglect the disciplines of logic and mathematics, for it is these sciences and these alone that can put him in right relation to what the Germans call the Gedankenwelt, to what the English call the world of ideas, the most potent of all the components of life and the world, for the universe itself is but an idea. I have not time to argue that one who hopes to acquire liberal education will not neglect the discipline of beauty, the most vital thing in the world, for it is beauty that makes life worth living and makes it possible. Indeed, if by some spiritual cataclysm all the beauty of nature and all the beauty of art and all the beauty of thought were to be suddenly blotted out, man would quickly perish through depression of spirit caused by the omnipresence of ugliness. Neither have I time to argue in this connection that one aspiring to a liberal education may not neglect the greatest of all the arts of men, the art which, at the first university meeting of the year President Wheeler so emphatically warned us not to neglect: the art, I mean, of human expression in living speech.

Is magnanimity alone the highest of ideals? It is not. Neither efficiency alone nor magnanimity alone is the highest of ideals. The supreme ideal is a union of the two. Its name is wisdom. But, you may wish to ask, why have I said nothing of righteousness? Have I said nothing of it? On the contrary I have been speaking of naught else. For in respect of this matter education, liberal education, agrees with Socrates: badness is a species of ignorance; the wise man, the wise woman, will be good.

INDEX OF AUTHORS